GERTRAUD MARIA DASCHIEL

Strategies to reduce friction losses and their implications for the energy efficient design of internal flow domains

SCHRIFTENREIHE DES INSTITUTS FÜR STRÖMUNGSMECHANIK
KARLSRUHER INSTITUT FÜR TECHNOLOGIE (KIT)

BAND 1

Strategies to reduce friction losses and their implications for the energy efficient design of internal flow domains

BY

GERTRAUD MARIA DASCHIEL

KIT Scientific Publishing

Dissertation, Karlsruher Institut für Technologie (KIT)
Fakultät für Maschinenbau, 2014

Tag der mündlichen Prüfung: 2. Oktober 2014
Referenten: Prof. Dr.-Ing. Bettina Frohnapfel
apl. Prof. Dr.-Ing. habil. Jovan Jovanović

Impressum

KIT Scientific Publishing

Karlsruher Institut für Technologie (KIT)
KIT Scientific Publishing
Straße am Forum 2
D-76131 Karlsruhe

www.ksp.kit.edu

Print on Demand 2015

ISSN 2199-8868
ISBN 978-3-7315-0291-3
DOI 10.5445/KSP/1000043770

Abstract

The aim of the present work is to develop a feasible strategy for the reduction of the energy that is required to transport fluids. In this respect, fully developed laminar and turbulent flows through straight ducts, which are found extensively in many practical applications, are considered. Based on the behavior of the flow in a specific flow regime, physical models are initially derived that lead to a reduction in the energy dissipated by the fluid motion, thus simultaneously reducing friction losses and providing energy savings. An attempt is made to attain the desired flow state due to appropriate designs for the cross section shape of the duct. Optimization of the pipe cross section geometry for the pure laminar regime can be accomplished analytically, resulting in the design of the preferred flow configurations.

For high Reynolds numbers, corresponding to the turbulent regime, the substantial complexity of the fluid motion produces an increase in the energy dissipation. Thus, ensuring laminar flow under conditions for which it is typically found to be turbulent represents the promising objective for achieving energy savings. This realization is undertaken using statistical tools that allow a mathematical description and correlation between the transitional and fully developed turbulent flows. These considerations lead to the conclusion that similar conditions in the statistical flow field are assumed to provoke the delay of transition to turbulence and the laminarisation of turbulent flow. Direct numerical simulations are performed with the intention of exploring the possibility of initiating these particular processes by forcing the flow structure towards the desired state through the pipe cross section geometry. For this purpose coexisting laminar and turbulent flow regions that appear in narrow corners of non-circular ducts are examined in great detail. Based on these findings, novel duct geometries are suggested with the final objective being the derivation of cross section configurations for which the laminar flow regime prevails over turbulence. The potential of various design aspects to achieve this goal is assessed and possible energy savings are quantified.

Zusammenfassung

Die vorliegende Arbeit zielt darauf ab technisch realisierbare Strategien zu entwickeln, welche es ermöglichen die Energie, die für den Transport von Fluiden benötigt wird, zu reduzieren. In diesem Kontext werden voll entwickelte, laminare und turbulente Strömungen durch gerade Rohre betrachtet, die in zahlreichen praktischen Anwendungen zu finden sind. Zunächst werden für die spezifischen Strömungseigenschaften eines Strömungsregimes physikalische Modelle hergeleitet, welche zur Reduktion der, durch die Fluidbewegung dissipierten Energie führen und damit gleichzeitig verringerte Reibungsverluste und Energieeinsparungen mit sich bringen. Darauf aufbauend wird die Herbeiführung dieser gewinnbringenden Strömungszustände durch die adäquate Gestaltung der Querschnittform eines Rohres in Angriff genommen.

Für rein laminare Strömungen ist die Optimierungsaufgabe für den Rohrquerschnitt analytisch behandelbar und bevorzugte Konfigurationen werden aufgezeigt.

Im Bereich hoher Reynolds Zahlen, in welchem das turbulente Strömungsregime vorherrscht, bedingt die hohe Komplexität der Fluidbewegung das Ansteigen der Energiedissipation. Aufgrund dieser Tatsache stellt die Laminarhaltung von Strömungen unter Bedingungen, unter denen sie typischerweise turbulent sind, das viel versprechende Ziel für das Erreichen von Energieeinsparungen dar. Durch den Einsatz von statistischen Werkzeugen, welche eine verwandte, mathematische Beschreibung von transitionalen und voll entwickelten turbulenten Strömungen erlauben, können Möglichkeiten zum Erreichen dieser Zielsetzung aufgezeigt werden. Basierend auf diesen Betrachtungen kann gefolgert werden, dass vergleichbare Eigenschaften des statistischen Strömungsfeldes zum einen zur Transitionsverzögerung und zum anderen zur Laminarisierung von turbulenten Strömungen führen. Die Möglichkeit, die gewünschte Struktur der Strömung und damit die relevanten Prozesse durch die Gestaltung der Rohrquerschnittgeometrie herbeizuführen, wird in direkten numerischen Simulationen untersucht. In diesem Zusammenhang werden koexistente laminare und turbulente Strömungsgebiete detailliert untersucht, wie sie in spitzwinkeligen Ecken von nicht kreisförmigen Rohren auftreten. Aufbauend auf diesen Erkenntnissen, werden neuartige Geometrien vorgeschlagen, welche zur Herleitung von Querschnittkonfigurationen dienen, für welche sich das laminare Strömungsregime gegenüber dem turbulenten durchsetzt. Das Potential verschiedener Designaspekte für das Erreichen dieser Zielsetzung wird bewertet und die damit einhergehenden Energieeinsparungen werden quantifiziert.

Contents

Abstract i

Zusammenfassung iii

1 Introduction 1

2 Conservation laws for the motion of incompressible Newtonian fluids 3

2.1 Conservation of mass and momentum 3
2.2 Conservation of kinetic energy 5

3 Objective and procedure 7

4 Frictional resistance and energy consumption of internal flows 9

4.1 General description of the flow cases investigated 9
4.2 Friction laws in laminar and turbulent flows 11
4.3 Entire energy consumption of the flow 13
4.4 Formulation of the optimization problem 14

5 Potentials and limits for energy savings in laminar flows 17

5.1 Mathematical description of the flow 17
5.2 Lower limit for the energy consumption 18
5.3 Surface structures leading to reduced friction losses of channel flow 19

6 Theoretical analysis of transitional and turbulent flows 25

6.1 Statistical description of fluid motion 25
6.1.1 Transport equations of turbulent flow 26
6.1.2 Basic equations for laminar flow with superimposed small disturbances 29
6.1.3 Interpretation of the statistical flow properties using invariant theory 30
6.2 Physical models for reduced friction losses of wall-bounded flows 33
6.2.1 Kinematic consideration of velocity fluctuations 33

6.2.2 Delay of transition to turbulence 34
6.2.3 Reduction of dissipation in turbulent flows 38
6.2.4 Review of investigations on friction drag reduction due to surface modifications 42
6.3 Conclusions for further investigation of internal flows 44

7 Presentation and evaluation of the numerical method 49
7.1 Numerical procedure for the calculation of turbulent flow in non-circular ducts 49
7.1.1 Numerical method 50
7.1.2 Spatial and temporal discretization 51
7.1.3 Initialization and evaluation of the turbulent flow field . . 52
7.2 DNS of turbulent flow in a square duct - comparison to literature data 53
7.2.1 Computation model 53
7.2.2 Special properties of the velocity field 57
7.2.3 The friction behavior 59
7.3 Simulation of the partial laminarization of turbulent flow 65

8 Verification of the mechanism leading to low energy dissipation in duct flows 71
8.1 Flow in a triangular duct with 11.5° apex angle 73
8.1.1 Properties of the coexisting laminar and turbulent flow field 73
8.1.2 Description of the transition process using the invariant representation of the stresses 81
8.2 Comparison of coexisting laminar and turbulent flow fields in related ducts 92
8.2.1 Characterization of stable laminar duct flow 92
8.2.2 Constraints for flow laminarization 94
8.2.3 Impact of corner designs on the transition scenarios 95

9 Impact of duct corners on turbulent flow 101
9.1 Flow behavior in classical non-circular ducts 102
9.1.1 Review of findings in the literature 102
9.1.2 Properties of the turbulent stresses 106
9.2 Design of novel duct geometries using corner effects 112
9.2.1 Impact on the turbulent stresses 113
9.2.2 Impact on the dissipative losses in the flow 123
9.2.3 Destabilization of the mean velocity profile 128

10 Quantification of potential energy savings 137

11 Conclusion and outlook 143

Bibliography 147

Nomenclature 157

List of Figures 163

List of Tables 169

A Validation data for model prediction of DR in laminar flow 171

Acknowledgements 173

1 Introduction

In recent last decades, engineers and scientists have been working intensively on possible ways to reduce the energy consumption of flow systems. These efforts are motivated by the increasing public awareness of the efficient usage of energy resources and by the continuous increase in energy costs. In general, various contributions to the energy dissipation in fluid flows can be distinguished. In non-reacting, isothermal flows of incompressible fluids, which are the focus of this work, pressure drag and friction drag are usually the most pronounced. In many internal flow applications, *e.g.* in pipelines, the contribution of friction losses to the total energy dissipation prevails. The reduction of these losses is expected to lead to significant savings and is investigated in the present work.

Previous and ongoing attempts towards reducing friction losses in internal flows have concentrated on the turbulent flow regime, which has led to many promising techniques being discovered. Passive methods have the longest tradition in this respect and the most significant example has been applied to oil pipelines since the 1970s. Small amounts of commercially available drag reduction additives (DRAs), consisting typically of long-chain polymers, result in the reduction of losses due to friction by up to 80% [3]. Manufacturers of DRAs promote their products with the slogan "Move more product with less energy and capital" [3]. In fact, in the case of the Alaska pipeline, in 2003, the use of DRAs facilitated an increase in the maximum daily throughput of more than 50% [2]. However, because of chemical and environmental reasons, drag reduction with DRAs is mainly only applicable to the transportation of crude oil, refined products and non-potable water [4]. In addition, the efficiency of DRAs generally suffers as a consequence of their degradation, which also depends on the working temperature and the flow speed.

The majority of drag reducing techniques focus on influencing the fluid friction due to specific features applied at the fluid–solid boundary. Changes in the wall topology represent another passive control technique and are considered as one of the earliest proposals in this respect. In a patent from 1937, Kramer [71] suggested arranging thin wires at a small distance above a flat surface. The wires are parallel to each other and aligned in the flow direction. In spite of the surface area increase, an overall drag reduction was reported. Since this pioneering work of Kramer, different types of structural geometries have been investigated, which are referred to as riblets. Extensive investigations dealing with the performance of these structures with respect to their design have been carried out. Following more or less a trial and error approach, structures were found that are capable of reducing the friction drag by up to 10% [11].

Moreover, active means for the reduction of friction drag in internal flows have been studied in the literature. Actuators at the wall enabling suction and blowing, or a defined movement are typical examples that are frequently discussed in this context. These techniques usually are optimized towards minimal friction losses within numerical parameter studies. By contrast to the presented passive control methods, active control actually requires the continuous input of energy. Thus, the reduction of friction drag does not necessarily result in energy savings. In fact, to date these techniques typically only lead to a small reduction in the total energy consumption [39].

Despite the drag reduction due to DRAs, the practical application of control methods is still an exception rather than the rule. One reason can be found in the fact that some of these strategies are only tested in simulations and are rather difficult to apply to practical applications. On the other hand, the implementation of drag reducing devices typically represents a significant effort in terms of costs and the compliance with safety standards. Thus, their practical relevance strongly depends on their efficiency operation.

In the present work, the question how friction losses in internal flows can be reduced is first approached theoretically. Strategies are derived for different flow regimes that can provoke a significant reduction of the friction drag. An attempt is made to implement the derived mechanisms by geometrical modifications in the flow domain. This method is applicable to a wide range of engineering applications including different kinds of fluids and is expected to be fairly robust with respect to the surrounding conditions. Most importantly, if this passive control method is followed, no energy input as such is required and reduced friction losses lead directly to a reduction in the energy consumption.

2 Conservation laws for the motion of incompressible Newtonian fluids

2.1 Conservation of mass and momentum

In a continuum mechanical frame it is generally accepted that the following set of partial differential equations describes the isothermal motion of fluids [106]:

$$\frac{\partial \rho}{\partial t} + \frac{\partial (\rho U_i)}{\partial x_i} = 0 \quad , \tag{2.1}$$

$$\rho \left[\frac{\partial U_i}{\partial t} + U_k \frac{\partial U_i}{\partial x_k} \right] = -\frac{\partial P}{\partial x_i} + \frac{\partial \tau_{ik}}{\partial x_k} + \rho f_i \quad . \tag{2.2}$$

In these equations U_i represents the velocity vector, P the pressure and ρ the density of the fluid. The term ρf_i describes the body force per unit volume and τ_{ik} is the stress tensor, having six independent components. The equations are given using index notation where $i, k, j = 1, 2, 3$ represent the components in a Cartesian coordinate system. Equation (2.1) ensures the conservation of mass and is referred to as the continuity equation while Equation (2.2) is derived from the requirement to conserve the momentum of a fluid volume element.
In order to close this system of equations, the number of unknown quantities has to be reduced. To do this, a deformation law has to be established in order to provide a formula to express the stress tensor, τ_{ik}, in terms of velocity derivatives. Throughout this work, the motion of incompressible (ρ = const) and Newtonian fluids is considered, which leads to [106]:

$$\tau_{ik} = \mu \left(\frac{\partial U_i}{\partial x_k} + \frac{\partial U_k}{\partial x_i} \right) \quad . \tag{2.3}$$

Here, μ represents the dynamic viscosity and can be transferred to the kinematic viscosity, $\nu = \mu/\rho$.

In the following, plane internal flows appearing in many fluid transport processes are considered for which the impact of body forces on the resulting velocity field typically vanishes. This conclusion is based on the comparison of the magnitude of individual terms in Equation (2.2) [28, 29]. Additionaly, it is supported by the agreement of solutions to the simplified equations with experimental findings [29]. However, it is pointed out that even if neglecting body forces is justified if the resulting velocity fields in many cases that are of practical importance are addressed, it does not apply generally to internal flows. For example, Draad and Nieuwstadt [28] found that Coriolis forces can affect the velocity profile that develops in a cylindrical pipe significantly. The effect is shown to strongly depend on the flow regime and the fluid properties and is only expected to be observed in a laminar flow of fluids with rather high kinematic viscosity.
Considering flow cases with a vanishing impact of body forces on the velocity field and using the deformation law in Equation (2.3), Equations (2.1) and (2.2) can be reformulated for an incompressible Newtonian fluid:

$$\frac{\partial U_i}{\partial x_i} = 0 \quad , \tag{2.4}$$

$$\frac{\partial U_i}{\partial t} + U_k \frac{\partial U_i}{\partial x_k} = -\frac{1}{\rho}\frac{\partial P}{\partial x_i} + \nu \frac{\partial^2 U_i}{\partial x_k \partial x_k} \quad , \tag{2.5}$$

while Equation (2.5) is referred to as the Navier–Stokes equations.
This set of differential equations can be transferred into a dimensionless form, where '$*$' denotes a dimensionless quantity. To do this, the dimensional quantities are substituted using constant properties that characterize the flow and are labeled with the index 'c':

$$U_i = U_c U_i^*; \quad x_i = L_c x_i^*; \quad t = t_c t^*; \quad \rho = \rho_c \rho^*; \quad P = \Delta P_c P^*; \quad \nu = \nu_c \nu^*.$$

Applying these relationships to Equations (2.4) and (2.5) yields [29]

$$\frac{\partial U_i^*}{\partial x_i^*} = 0 \quad , \tag{2.6}$$

$$St \frac{\partial U_i^*}{\partial t^*} + U_k^* \frac{\partial U_i^*}{\partial x_k^*} = -Eu \frac{1}{\rho^*}\frac{\partial P^*}{\partial x_i^*} + \frac{1}{Re}\nu^* \frac{\partial^2 U_i^*}{\partial x_k^* \partial x_k^*} \quad , \tag{2.7}$$

and the characteristic quantities are used to define the following dimensionless numbers,

$$St = \frac{L_c}{U_c t_c}; \quad Eu = \frac{\Delta P_c}{\rho_c U_c^2}; \quad Re = \frac{U_c L_c}{\nu_c} = \frac{\rho_c U_c^2}{\tau_c} \quad . \tag{2.8}$$

These numbers are referred to as the Strouhal number, St, Euler number, Eu, and Reynolds number, Re, and describe the relevant dynamic features of the flow. Thus, flows with similar geometrical properties and boundary conditions are dynamically similar as well, if the dimensionless numbers are kept constant [29].
By setting the dimensionless numbers equal to one, relationships for the characteristic velocity, length and time scale can be derived, as demonstrated by Durst [29]. The resulting scaling will be applied to the wall bounded flows considered further on in this work. For this class of flows of fluids with the density ρ and the kinematic viscosity ν, the characteristic velocity scale reads as

$$U_c = u_\tau = \sqrt{\tau_w/\rho} \quad , \tag{2.9}$$

where τ_w is the characteristic shear stress which arises at the wall. The characteristic behavior close to the wall also determines the temporal and spatial scales of the flow:

$$t_c = \frac{\nu}{u_\tau^2} \quad , \qquad L_c = \frac{\nu}{u_\tau} \quad . \tag{2.10}$$

2.2 Conservation of kinetic energy

The major objective of this work is based on the evaluation of possible energy savings due to geometrical modifications of the fluid–solid boundary. Thus, the conservation equation of kinetic energy, which is derived from the momentum equation (2.2) is of central importance [54]:

$$\underbrace{\frac{1}{2}\frac{\partial U_i U_i}{\partial t}}_{I} = \underbrace{-\frac{\partial}{\partial x_k} U_k \left(\frac{P}{\rho} + \frac{U_i U_i}{2}\right)}_{II} + \underbrace{\nu \frac{\partial}{\partial x_k}\left[U_i\left(\frac{\partial U_k}{\partial x_i} + \frac{\partial U_i}{\partial x_k}\right)\right]}_{III} \underbrace{- \nu \left(\frac{\partial U_k}{\partial x_i} + \frac{\partial U_i}{\partial x_k}\right)\frac{\partial U_i}{\partial x_k}}_{IV} \quad . \tag{2.11}$$

The individual terms in Equation (2.11) can be physically interpreted as follows (where all descriptions are defined as per unit mass and time):

- I: local change of kinetic energy
- II: change in convective transport of the total energy
- III: work done by the viscous stresses
- IV: energy dissipation, Φ.

The dissipation function,

$$\Phi = \nu \left(\frac{\partial U_k}{\partial x_i} + \frac{\partial U_i}{\partial x_k} \right) \frac{\partial U_i}{\partial x_k} \quad , \tag{2.12}$$

is a measure for the energy that is converted per unit mass and time into heat in an irreversible fashion and therefore always has to be positive [104]. Obviously, changes in this process have a direct impact on the energy that has to be applied to maintain the movement in a certain flow system.

3 Objective and procedure

Depending on their appearance, wall-bounded flows can generally be classified into two major regimes: laminar flows show a high level of regularity, which is progressively lost, if the flow turns, during the so-called transition process, into the turbulent state. This process was found to be accompanied by an increase in the dimensionless Reynolds number, where the critical Reynolds number is defined to describe the border between laminarity and turbulence. The substantial differences in the flow field are reflected by the magnitude of the friction losses on the walls, which are significantly higher in turbulent than in laminar flows. In the case of fully developed, plane flow situations, which will be treated throughout this work, these friction losses alone are responsible for the energy dissipation that occurs, which is often referred to as flow resistance. Thus the energy that has to be applied in order to maintain the movement in a system, *e.g.* the volume flow rate pumped through a pipe, is only caused by the fluid friction on the solid wall and consequently is considerably smaller for laminar than for turbulent flows.

This work focuses on studying the possibility of reducing the flow resistance of internal, fully developed flows due to geometrical modifications of the fluid–solid boundary. In this context, the similarities and differences between the strategies leading to a gainful interface design in the different flow regimes are emphasized. In a first stage, the friction behavior of pure laminar flow is considered, which typically is supposed to lead to minimum losses [12]. Hence, the question that has be discussed is, if and how this friction behavior can be influenced due to geometrical modifications in the flow domain.

For increasing Reynolds numbers disturbances in the initially stable, laminar flow state begin to be amplified, which finally leads to the breakdown to turbulence. Previous investigations have shown that the flow field during this process and at the final turbulent state can be described using a statistical approach, which splits the appearing flow quantities into a mean and a fluctuating part [67, 103]. Based on this approach, conditions can be derived analytically that lead to a vanishing impact of the fluctuating quantities on the flow, resulting in the maintenance of laminar flow even at high Reynolds numbers and the laminarization of turbulent flow at the same time [65, 66].

The goal of this work is to derive duct geometries that force the flow to approach these conditions, which are expected to lead to a significant reduction in friction drag. In this context a series of direct numerical simulations of duct flows is performed, which initially require a detailed evaluation of the accuracy in the numerical procedure.

After doing this, a triangular duct having a small apex angle is investigated. For this duct geometry it was observed experimentally that a laminar and a turbulent region can coexist for a certain range of Reynolds numbers, for which the flow can generally be expected to be fully turbulent [30]. This specific flow field shows that the laminarization of turbulent flow due to appropriate geometrical properties, namely corner regions, is possible. In addition, further investigations aimed to see whether the laminarization and transition processes in the triangular duct actually follow the theoretically derived behavior. This question is of particular interest as a positive result suggests that properly designed duct geometries can, in agreement with the theoretical prediction, result in maintenance of laminar flow and laminarization of turbulent flow simultaneously. This study is complemented by the investigation of further internal flows that lead to coexisting laminar and turbulent regions in order to generalize the observations made for the triangular duct.

Initiated by these results, a systematic investigation of the impact of corner regions on turbulent flow, considering different corner angles and wall curvatures, is performed. In doing so, the ability of a certain geometrical property to locally force the flow in the theoretically suggested state can be analyzed. In conclusion, several duct geometries can be derived that are supposed to provide this state in a large part of the flow domain. For a duct geometry, which is promising in this sense, the ability to finally laminarize turbulent flow on the one hand and to maintain laminar flow on the other is analysed.

In addition to studies dealing with the influence of the duct geometry on the resulting friction, the important question, under which condition these insights can lead to benefits in engineering practice, is finally discussed.

4 Frictional resistance and energy consumption of internal flows

Within this chapter, the common set-up of the flow cases investigated throughout this work is introduced. The laws describing the frictional resistance of these flows are given and connected to their overall energy consumption. Based on these relationships, the optimization problem addressed within this work is formulated.

4.1 General description of the flow cases investigated

Within this work, pressure driven flows in ducts that have an arbitrary cross section shape are considered. The straight and plane flow domains are bounded by fixed, solid walls and the cross section shape does not vary in the streamwise x_1-direction, in the example as shown in Figure 4.1. The area of the cross section is referred to as A_{cs} and A_w denotes the surface area of the wall on which the flow obeys the no-slip boundary condition. The flow through these ducts is governed by the continuity and the Navier–Stokes equation given in (2.4) and (2.5).

These flows generally can appear in two different flow regimes, laminar and turbulent. While the laminar flow is considered to be stationary, the nature of turbulence leads to instantaneous velocity and pressure fields that are fluctuating in time and space. However, time averaging enables the description of the flow field by mean quantities, $\overline{U}_i$ and $\overline{P}$. The mean quantities are supposed to be stationary for time periods that extend the averaging time.[1]

Further, the position $x_1 = 0$ is assumed to be located at some distance from the entrance, where the laminar and the time-averaged turbulent velocity profile have reached constant shapes that do not vary along L_{x_1}.

[1] In this Chapter, only time-averaged quantities are used to describe the properties of turbulent flow, which are also referred to as mean quantities. For clarity, this fact is further illustrated in equations but not repeated explicitly in the text.

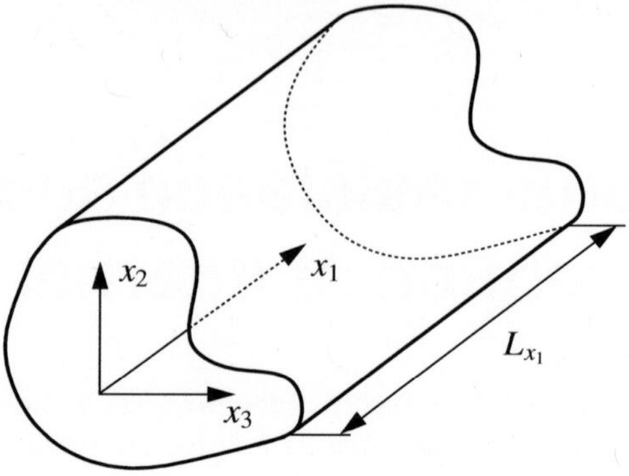

Figure 4.1: Straight duct with arbitrary cross section shape that is homogeneous in the streamwise x_1-direction. L_{x_1} denotes the length of the considered flow domain.

The consideration of time-averaged turbulent quantities enables the derivation of common relations for the frictional resistance and the entire energy dissipation in both flow regimes. These relationships will be discussed in the following sections.

4.2 Friction laws in laminar and turbulent flows

The flow resistance of stationary and fully-developed duct flows is frequently studied in the literature. Initial investigations mainly focus the determination of the flow resistance using the integral balance of the forces acting in the streamwise direction. The major findings are summarized briefly, according to the description by Schlichting [106].
Volume integration of the momentum balance in the x_1-direction leads to the equilibrium of forces in the considered flow configuration. It is observed that the difference between the inlet and outlet pressure forces are balanced by the shear forces which are acting on the duct wall:

$$\Delta P\, A_{cs} = \tau_w\, A_w \quad . \tag{4.1}$$

The pressure difference ΔP is evaluated according to

$$\Delta P = P(x_1 = 0) - P(x_1 = L_{x_1}) \quad \text{or} \quad \Delta P = \overline{P}(x_1 = 0) - \overline{P}(x_1 = L_{x_1}) \quad , \tag{4.2}$$

where τ_w corresponds to the surface averaged shear stress in the x_1-direction appearing at the duct wall. The local wall shear stress is defined as the gradient of the streamwise velocity component in wall normal direction and is not constant along the perimeter of non-circular ducts.
In engineering practice, the dimensionless friction factor, f, is typically used to characterize the frictional resistance of internal flows. f is a measure of the surface averaged shear stress at the wall and is defined as

$$f = \tau_w \frac{8}{\rho U_b^2} = \frac{\Delta P A_{cs}}{A_w} \frac{8}{\rho U_b^2} \quad , \tag{4.3}$$

where U_b is the volume averaged streamwise velocity, which is referred to as bulk velocity,

$$U_b = \frac{1}{V} \int_\Omega U_1 \, \mathrm{d}V \quad \text{or} \quad U_b = \frac{1}{V} \int_\Omega \overline{U}_1 \, \mathrm{d}V \quad . \tag{4.4}$$

The operator $\int_\Omega () \mathrm{d}V$ denotes the volume integral over the entire domain Ω. In Equation (4.3), the friction factor is also expressed in terms of the pressure difference arising between the inlet and the outlet of the internal flow domain. This relationship is derived from the balance of forces in Equation (4.1). It should be noted that in German literature f is it often referred to as λ, *e.g.* in the book by Schlichting [106].
The dimensionless form of the Navier–Stokes equations (2.7) shows that the considered stationary flows are characterized by the Reynolds number and the

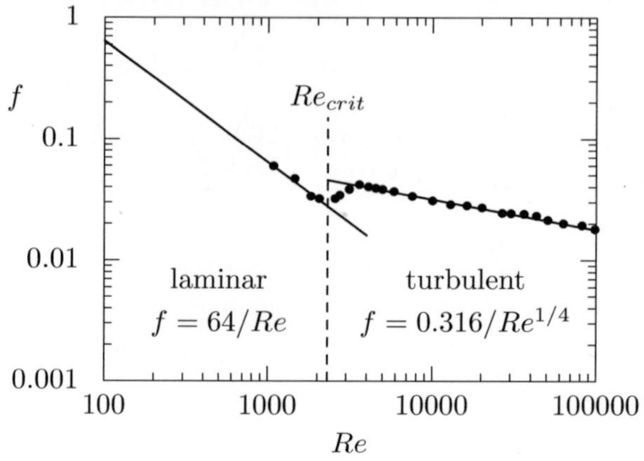

Figure 4.2: Dependence of the friction factor on the Reynolds number for laminar and turbulent flow through a smooth cylindrical pipe. Data presented by Schlichting [106] has been replotted.

Euler number. A comparison of Equations (2.8) and (4.3) yields the proportionality between the previously introduced friction factor f and the Euler number. Thus, it can be concluded, for both characteristic numbers, that they are only a function of the Reynolds number [29]. However, for the present flow cases f is preferably used for dimensionless considerations. In the following, the friction laws, which are based on the dependence of f on Re, are first discussed for cylindrical pipes before ducts of arbitrary cross section shapes are addressed.
In laminar flow, the friction behavior of cylindrical pipe flow obeys the analytically derived relationship $f = 64/Re$, were the Re is based on the pipe diameter and the bulk velocity. In turbulent flow, the following empirical relation was found to describe the friction factor for flow through a cylindrical pipe with smooth walls:

$$f = 0.316/Re^{1/4} \quad . \tag{4.5}$$

This equation is referred to as the Blasius correlation.
The relations for laminar and turbulent flow through a cylindrical pipe are plotted in Figure 4.2. The critical Reynolds number, $Re_{crit} = 2300$, after which sustained turbulent flow can be observed, is also sketched. Experimental data for the flow through a smooth cylindrical pipe [106] agree with the laminar solution and support the correlation suggested for the turbulent regime.

The Blasius correlation was found to also describe the friction of turbulent flow through different ducts with non-circular cross sections if the hydraulic diameter is used as the characteristic length scale. The definition of the hydraulic diameter is based on the ideas of v. Mises [84]:

$$D_h = \frac{4\,A_{cs}}{C} \quad , \tag{4.6}$$

where $C = A_w/L_{x_1}$ is the perimeter of the cross section. The Reynolds number based on D_h is typically referred to as the hydraulic Reynolds number, Re_h. In contrast to the findings for turbulence, the friction law is not uniform if laminar flow through non-circular ducts is considered. In this case, the friction factor obeys the relationship

$$f = a/Re_h \quad , \tag{4.7}$$

and the constant a is unique for a certain duct geometry.

4.3 Entire energy consumption of the flow

Frohnapfel [38] analysed the energy balance for the class of flows considered here. The corresponding derivations focus on turbulent flow but can also be assigned to the laminar regime. Starting from the conservation equation of kinetic energy (2.11) in the streamwise direction and integrating over the volume of the domain leads to the following equilibrium:

$$-\frac{1}{\rho}\frac{\mathrm{d}P}{\mathrm{d}x_1}\int_\Omega U_1\,\mathrm{d}V = \int_\Omega \Phi\,\mathrm{d}V \quad \text{or} \quad -\frac{1}{\rho}\frac{\partial \overline{P}}{\partial x_1}\int_\Omega \overline{U}_1\,\mathrm{d}V = \int_\Omega \Phi\,\mathrm{d}V \quad . \tag{4.8}$$

Using results from the balance of forces in Equations (4.1) and (4.3), the last relationships can be reformulated:

$$\langle\Phi\rangle = \rho\int_V \Phi\,\mathrm{d}V = \Delta P U_b A_{cs} = A_w \tau_w U_b \quad , \tag{4.9}$$

It can be concluded that the entire energy dissipation rate of the working fluid, $\langle\Phi\rangle$, is balanced by the pressure difference that is required to drive a defined flow rate, $\dot{V} = U_b A_{cs}$, in both flow regimes. Further, $\langle\Phi\rangle$ is directly connected to the frictional resistance induced by the duct wall.

In practice, a pump has to permanently supply energy in order to overcome the irreversible losses and to maintain the movement of the fluid in an internal flow domain. This expenditure is typically expressed in terms of the pumping power,

P_P. The classical definition of P_P is given in the following equation. Additionally, applying Equation (4.9) yields the balance of P_P and the absolute value of the entire dissipation rate of the working fluid:

$$P_P = \Delta P \, \dot{V} = \langle \Phi \rangle \quad . \tag{4.10}$$

4.4 Formulation of the optimization problem

The present investigation addresses the reduction in the energy that is required to drive the previously described internal flows. The aim is to achieve this objective through the modification of the flow by a control method that is acting on the wall. In principle, different control techniques are available for this purpose.
Active methods are found to yield the reduction of friction forces acting on the walls which, in theory, is promising. However, it has to be considered that additional energy input is required to drive the control, which contributes to the total energy consumption of the flow system. Thus, reduced friction forces do not automatically lead to overall energy savings.
In contrast, for all passive approaches, the entire energy consumption of the flow system consists of energy dissipated by the working fluid itself and is directly linked to the friction losses on the wall. Mathematically, the considered cost function can be formulated as

$$J(\Omega) = \rho \int_{\Omega} \Phi \, \mathrm{d}V = \langle \Phi \rangle \quad . \tag{4.11}$$

The focus is to reduce $\langle \Phi \rangle$ compared with presently applied configurations due to the appropriate design of the duct cross section shape. In doing so, the cross section area, the bulk velocity and the volume of the pipe are kept constant. At best, a cross section shape is found which minimizes $J(\Omega)$.
The performance of geometrical variations can be evaluated according to

$$\Delta\langle \Phi \rangle = 1 - \frac{\langle \Phi \rangle_{\text{modified flow domain}}}{\langle \Phi \rangle_{\text{initial flow domain}}} \quad , \tag{4.12}$$

while $\Delta\langle \Phi \rangle$ denotes energy savings. In the literature, the reduction of the friction drag, *DR*, is also used to measure the performance of flow control methods:

$$DR = 1 - \frac{(A_w \tau_w)_{\text{modified flow domain}}}{(A_w \tau_w)_{\text{initial flow domain}}} \quad . \tag{4.13}$$

According to Equation (4.9), the measures are identical for the addressed flows: $\Delta\langle \Phi \rangle = DR$.

The geometrical modifications that lead to energy savings, based on theoretical considerations, are the focus of the discussions in the following chapters. Even if the entire energy consumption of the laminar and time-averaged turbulent flows obey similar expressions, the physical mechanisms describing the motion in the flow regimes are different. The specific properties of laminar and turbulent motion are used to derive individual models that lead to the reduction of $\langle \Phi \rangle$.

5 Potentials and limits for energy savings in laminar flows

5.1 Mathematical description of the flow

In the following, the conservation equations given in Chapter 2 are formulated for laminar internal flow. For this purpose, the stationary and fully developed flow case that was introduced in Chapter 4 is considered, which implies $\partial U_i/\partial t = 0$ and $\partial U_i/\partial x_1 = 0$. The continuity equation for this type of flow reads

$$\frac{\partial U_2}{\partial x_2} + \frac{\partial U_3}{\partial x_3} = 0 \quad , \tag{5.1}$$

while the Navier–Stokes equations give rise to

$$i = 1: \; U_2\frac{\partial U_1}{\partial x_2} + U_3\frac{\partial U_1}{\partial x_3} = -\frac{1}{\rho}\frac{\partial P}{\partial x_1} + \nu\frac{\partial^2 U_1}{\partial x_2 \partial x_2} + \nu\frac{\partial^2 U_1}{\partial x_3 \partial x_3} \quad , \tag{5.2a}$$

$$i = 2: \; U_2\frac{\partial U_2}{\partial x_2} + U_3\frac{\partial U_2}{\partial x_3} = -\frac{1}{\rho}\frac{\partial P}{\partial x_2} + \nu\frac{\partial^2 U_2}{\partial x_2 \partial x_2} + \nu\frac{\partial^2 U_2}{\partial x_3 \partial x_3} \quad , \tag{5.2b}$$

$$i = 3: \; U_2\frac{\partial U_3}{\partial x_2} + U_3\frac{\partial U_3}{\partial x_3} = -\frac{1}{\rho}\frac{\partial P}{\partial x_3} + \nu\frac{\partial^2 U_3}{\partial x_2 \partial x_2} + \nu\frac{\partial^2 U_3}{\partial x_3 \partial x_3} \quad . \tag{5.2c}$$

Following Equation (2.12), the dissipation function of the given flow in the streamwise direction ($i = 1$) leads to:

$$\Phi = \nu\left(\frac{\partial U_1}{\partial x_2}\right)^2 + \nu\left(\frac{\partial U_1}{\partial x_3}\right)^2 \quad . \tag{5.3}$$

Obviously, the cross flow components, U_2 and U_3, do not affect the energy dissipation, which has to be compensated to drive the flow. Actually, the velocity field of the considered flow through a straight duct having an arbitrary cross section shape is typically assumed to be purely axial, resulting in $U_2 = U_3 = 0$. From this assumption, a mathematical solution to Equations (5.1) and (5.2) is obtained. This solution is not proven to be unique but it is found to lead to a good agreement of the corresponding analytical results with measurements and is generally accepted for the description of the present flow situations [114]. According to these conditions, the mass conservation of the flow is fulfilled by definition and the Navier–Stokes equations (2.5) simplify significantly resulting in the so-called Poisson equation:

$$0 = -\frac{1}{\rho}\frac{\mathrm{d}P}{\mathrm{d}x_1} + \nu\left(\frac{\partial^2 U_1}{\partial x_2 \partial x_2} + \frac{\partial^2 U_1}{\partial x_3 \partial x_3}\right) \quad , \tag{5.4}$$

where a uniform pressure drop, $\mathrm{d}P/\mathrm{d}x_1$, impressed along its length, drives the flow. Obviously, the non-linear inertia term arising at the left hand side of the Navier–Stokes equations does not influence the flow situation described by Equation (5.4). Thus, the considered flow fields are dominated by viscous effects.

5.2 Lower limit for the energy consumption

In stationary, laminar and fully developed flows, the complexity of the conservation equations reduces significantly compared with the general form presented in Chapter 2. This fact allows an analytical solution of the shape optimization problem formulated in Equation (4.11). As a result, Schulz [107] finds the cylindrical pipe to lead to the lowest dissipation rate, $\langle\Phi\rangle$, when transporting a given flow rate through pipes having a constant volume and cross section area. According to Equation (4.9), the minimization of the dissipation rate and consequently also of the entire energy consumption due to the circular cross section shape is directly connected to the minimal friction forces acting on the pipe wall.

Further, the question of whether the losses of laminar internal flows can be reduced using active control techniques is discussed in literature. Using surface blowing and suction in the form of an upstream traveling wave, Min *et al.* [83] find that the frictional resistance of an internal flow can be sustained below that corresponding to the laminar regime. The resulting friction force on the wall is referred to as sub-laminar drag. However, theoretical analysis by Bewley [12] show that even if the friction force acting on the wall of internal flows can be reduced due to active methods, the lowest entire dissipation rate of the flow system belongs to uncontrolled laminar flow. The energy that is required to drive any active control is shown to be higher than the potential benefits that can be gained in this flow regime. This finding generalizes the optimality of the plain cylindrical pipe in laminar flows if the energy consumption is to be minimized.

The optimum represented by the cylindrical pipe might only be in question, if additional constraints come into play. The shape optimization problem solved by Schulz [107] then has to be reformulated according to these constraints, which can

be various but preferentially have practical relevance. A corresponding problem formulation is treated in the following discussion.
In engineering practice, a series of applications exist where fluid is transported in a flow domain that is restricted in its height rather than its width. These flow situations are often referred to as channel flows and are found, for example, in cooling ducts of numerous electronic devices and heat exchangers. For channel flow it is not proven that flat walls lead to minimal losses. The possibility to reduce the energy dissipation in a channel flow due to two-dimensional surface structures that are aligned with the main flow is discussed in the following section.

5.3 Surface structures leading to reduced friction losses of channel flow

The application of different shapes of rib-like surface structures oriented parallel to the flow direction as drag reducing devices in laminar channel flows have been investigated several times in recent decades. While it is well known that so-called riblets can lead to a significant drag reduction of up to 10% in turbulent flows [21], a similar result is not obtained for laminar flows (see Equation (4.13) for the definition of DR). For this case, riblets are actually found to increase the flow resistance compared with unstructured channel walls, leading to negative values of DR [20, 26, 85]. It should be noted, that the corresponding evaluation here and in the following discussions is based on keeping an equal cross section area and bulk velocity and therefore a constant flow rate in the structured and unstructured configuration. A less conservative attempt is followed within other works, as discussed in reference [26].
Firstly, the observed increase in flow resistance due to riblets in laminar flow is not too surprising since the mechanisms leading to benefits in turbulence were found in the interaction with specific properties of the flow in this state. Thus, there is no physical indication as to why this sort of surface modifications should also cause a reduced flow resistance in laminar flows, where energy is dissipated due to viscous effects only. However, using a variational principle for the surface shape, Pironneau and Arumugam [98] were able to show analytically that in the laminar case benefits in the viscous dissipation can be expected if the riblets exceed a certain width, $2l$, which scales with the mean channel height, $2L$, as follows (the definitions of the variables are given in Figure 5.1):

$$l/L > \pi/z \quad \text{where} \quad z \approx 1.2 \text{ is the solution of } 1 - z\tanh z = 0 \quad . \tag{5.5}$$

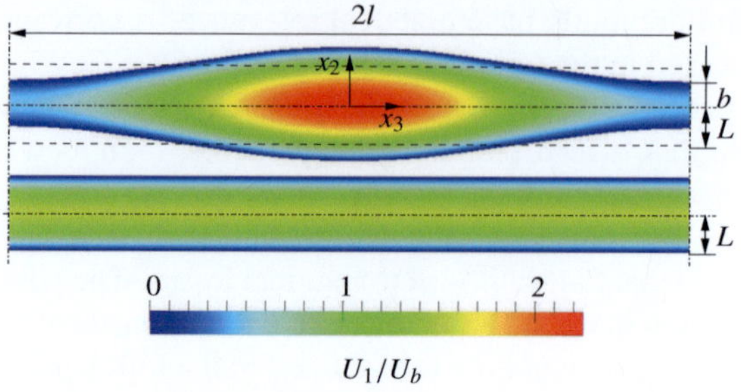

Figure 5.1: Numerically predicted laminar flow through a flat channel (bottom) and channel with wide surface structures (top). The boundary shape is defined by the function $x_2 = \pm((a/2)\cos(\pi x_3/l) + 2b)$ with $a/2 = L - b$ (a is the amplitude of the wave), here $l/L = 8.3$ and $a/(a+b) = 0.59$. Both channels have the same cross section area, which is indicated by the dashed side walls in the upper figure.

Performing a numerical shape optimization for certain l/L that obeys the above condition, Pironneau and Arumugam [98] find surface structures with smooth contours to reduce friction losses compared with the flat channel. In the following, this type of structure is further investigated by approximating the boundary found by Pironneau and Arumugam through a trigonometric function of the form

$$x_2 = \pm((a/2)\cos(\pi x_3/l) + 2b) \quad , \tag{5.6}$$

with $a/2 = L - b$ (a is the amplitude of the wave) [22]. In Figure 5.1 a typical structure geometry is shown together with the variables that are used for its description. The impact of the surface structure on the velocity field is also illustrated as a contour plot and is compared with the flat reference channel. The fully developed, stationary flow fields having the same bulk velocity are evaluated solving Equations (5.1) and (5.2) numerically and symmetric boundary conditions are applied to the lateral boundaries. This type of boundary condition is used to model a domain of infinite width. Thus, instead of considering the cross section area, the mean height of both channels is fixed to be identical, as indicated by the dashed lines in the upper figure. According to Equations (4.12) and (4.13) the drag and the energy consumption are reduced by 16% in the example shown in Figure 5.1, where $l/L = 8.3$ and $a/(a+b) = 0.59$.
In the following, the influence of the all parameters (l, a and b) determining the trigonometric structure on the frictional resistance is investigated. For this purpose, an analogy between structural mechanics and fluid mechanics can be used. Owing to the analogy between the torsion of beams and fully developed laminar flow

in ducts (the governing equation in both cases is Poisson's equation (5.4)), the pressure drop, and thus the skin friction drag, arising from the curved structures can be estimated by applying Saint Venant's principle [9]. The drag reduction is again evaluated by comparison with the result for the flat channel having the same mean height, $2L$, the same width, $2l$, and the same length, L_{x_1}, which obeys the analytical relation [29]

$$A_w \tau_w = \frac{12\,\mu\,U_b\,l\,L_{x_1}}{L} = \frac{3\,\mu\,\dot{V}\,L_{x_1}}{L^2} \quad . \tag{5.7}$$

The results for the investigated parameter range is shown on the left plot in Figure 5.2. However, it has to be considered that the model introduced by Bahrami [9] represents the flow through closed ducts, namely ducts that would be pictured by no-slip boundary conditions on the lateral sides of the domain shown in Figure 5.1. In order to consider the inaccuracies resulting from the model prediction, the drag reduction arising at certain parameter combinations in Figure 5.2 is compared with results of Pironneau and Arumugam [98] and our own numerical results, which were obtained as described previously. The corresponding values are tabulated in Appendix A. Overall, a good agreement is observed suggesting a small impact of the lateral walls on the results in the entire parameter range. In particular, in cases were the ratio l/b reaches high values, the model results are supposed to allow quantitative conclusions about the energy saving that can be obtained by replacing flat channel walls with the proposed surface structures (see Appendix A).
In Figure 5.2, the drag is found to be reduced compared with unstructured walls within the entire parameter range investigated. This result is expected since l/L obeys the theoretical condition for energy savings given in Equation (5.5). In general, it is observed that DR increases when moving to the upper right corner of the contour map representing wide structures that are periodically clamped such that b vanishes, *i.e.* the structured channel turns into a sequence of ducts. For this parameter range, a high accuracy of the model prediction is expected. The results indicate an asymptotic behavior in the drag reduction that can be achieved for large l/L, *i.e.* $DR_{\max} \approx 50\%$.
Results for triangular surface structures with sharp corners are shown on the right plot in Figure 5.2. Generally, the influence of variations in the parameters describing the structure are similar, as observed for the curved surface shapes. However, the drag reduction that can be achieved with a triangular surface structure is generally smaller and reaches a maximum of about 40%. The angeled shapes considered resemble classical riblets that are used as drag reducing devices in turbulent flow [21]. However, the width of the riblets studied in the literature is significantly smaller than the height of the channel. Thus, in the context of the present analysis, Choi's [20] observation of increased drag in laminar flow over

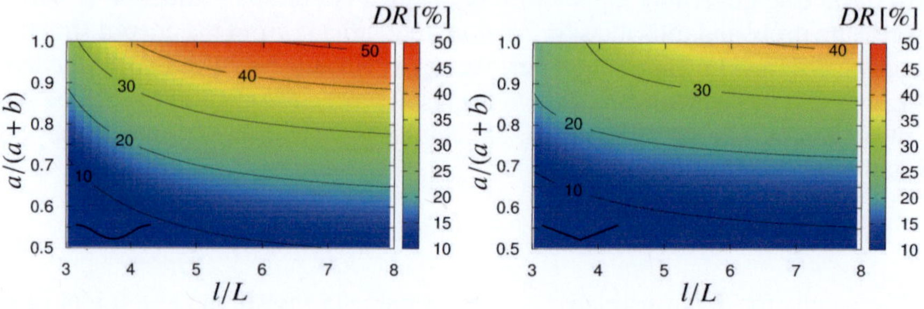

Figure 5.2: Contour plot of the drag reduction that can be achieved with structured walls compared with a flat reference channel depending on the parameters a, b and l that describe the structure geometry. Left plot: curved surface structure. Right plot: riblet-like triangular surface structure.

these riblets is no longer surprising and can be explained: in order to achieve energy savings in the laminar regime for different types of structure shapes, the ratio l/L has to exceed the critical border derived by Pironneau and Arumugam [98].

To summarize, the possibility of achieving energy savings in stationary and fully developed laminar flow due to geometrical variations in the boundary is limited. Schulz' [107] mathematical solution of the optimization problem shows that the circular pipe is optimal in this respect. However, in the case of channel flow, surface structures that are wide compared with the channel height can be shown to lead to reduced friction losses in comparison with flat walls. These surface structures provoke the increase and decrease of the channel height in a periodic fashion. The channel height generally has a strong impact on the friction forces acting on the walls if a fixed amount of fluid is to be transported (see Equation (5.7)). Obviously, this effect compensates the increase that is associated with the wetted surface area. The highest value for DR in this sense is found if the channel turns to a sequence of individual ducts. This observation indicates tendencies towards the general optimum, namely the cylindrical pipe.

In the following discussions, flows at higher Reynolds numbers, which become unstable for disturbances and tend to the turbulent state, are investigated. Again, internal flows in domains that are homogeneous in the streamwise direction are focused on. Similar to laminar flow, the height of a channel significantly influences the arising drag and the previously discussed surface structures also can lead to energy savings [22]. However, in contrast to the laminar regime, the energy dissipation in these cases is not just due to viscous effects and the universal optimality of the circular pipe can no longer be proven. Thus, tackling the general form of the optimization problem formulated in Equation (4.11) is felt to represent a realistic challenge.

6 Theoretical analysis of transitional and turbulent flows

In the previous chapter, the investigation of internal flows was restricted to the stationary, fully developed, laminar flow regime. It was shown that the high order of regularity in laminar flow is reflected by major simplifications in the transport equations. The flow behavior was observed to be dominated by viscous effects independent of the Reynolds number. However, as the Reynolds number exceeds a certain value, the non-linear term in the Navier–Stokes equation affects the fluid motion. In practice, the flow becomes sensitive to disturbances that originate, for example, from the surroundings or from corrugations in the domain boundary. The destabilization of the flow due to the disturbances finally results in the transition to turbulence.

The reduction of the energy that is required to drive statistically stationary, fully developed, internal flows at higher Reynolds numbers will now be considered. The energy balance of this type of flow was discussed in Chapter 4. The example for cylindrical pipe flow in Figure 4.2 illustrates that the friction factor, and also conclusively the energy consumption, of laminar flow at a given Reynolds number $Re > Re_{crit}$ is significantly lower than in turbulent flow. Thus, keeping the flow laminar at conditions where it usually turns to turbulence is supposed to be a straightforward goal for control strategies.

In order to achieve the present objective, a statistical description of the flow is introduced. The statistical framework forms the basis for the theoretical derivation of mechanisms in the flow leading to reduced dissipative losses in the turbulent regime, and at best, provokes the persistance of the laminar state.

6.1 Statistical description of fluid motion

The physical state of the flow at higher Reynolds numbers suggests the usage of statistical tools for its description. These tools will be introduced next and are first applied to fully turbulent flow fields for which the procedure is well established. The equations governing the motion and the energy conservation of the flow are given. The statistical tools are further used to describe laminar flows that are affected by disturbances and eventually undergo the transition to turbulence. Thus, the statistical framework allows a related description of the flow in the different regimes. This fact forms the basis for the derivation of common conditions leading to energy savings at higher Reynolds numbers.

6.1.1 Transport equations of turbulent flow

A turbulent flow field is generally characterized by quasi-random, time-dependent and three-dimensional variations of all flow quantities that are spread over a wide range of scales [63]. The largest scales in internal flows are determined by the flow domain, while the smallest scales represent the dissipation range. The motion of turbulent flow is also described by the continuity equation (2.4) and the Navier–Stokes equations (2.5). However, owing to the quasi-chaotic appearance of turbulent flow, it is most suitable to use statistical tools for its further description. In order to allow a better understanding of turbulent motion, Reynolds [103] proposes decomposing an instantaneous flow quantity, G, into the time-averaged mean, $\overline{G}$, and the time-dependent fluctuation, g:

$$G = \overline{G} + g \quad . \tag{6.1}$$

$\overline{G}$ is defined as

$$\overline{G}\,(x_1, x_2, x_3) = \lim_{T\to\infty} \frac{1}{T} \int_0^T G\,(x_1, x_2, x_3)\,\mathrm{d}t \quad , \tag{6.2}$$

where T is a sufficiently long time interval compared with the characteristic time periods of the fluctuation g.
Applying the decomposition proposed by Reynolds to the flow quantities leads to

$$U_i = \overline{U}_i + u_i, \quad P = \overline{P} + p \quad . \tag{6.3}$$

This splitting of the velocity components and the pressure is inserted into the continuity equation (2.4) and the Navier–Stokes equations (2.5). After temporal averaging of the resulting equations and considering averaging rules, the following conservation laws for the mean turbulent flow are obtained [63]:

$$\frac{\partial \overline{U}_i}{\partial x_i} = 0 \quad , \tag{6.4}$$

$$\frac{\partial \overline{U}_i}{\partial t} + \overline{U}_k \frac{\partial \overline{U}_i}{\partial x_k} + \frac{\partial \overline{u_i u_k}}{\partial x_k} = -\frac{1}{\rho}\frac{\partial \overline{P}}{\partial x_i} + \nu \frac{\partial^2 \overline{U}_i}{\partial x_k \partial x_k} \quad . \tag{6.5}$$

The motion of the mean flow is characterized by the interaction of the fluctuating quantities, $\overline{u_i u_k}$. This symmetric tensor of rank two contains six additional unknowns that lead to an unclosed system, which is referred to as the turbulent closure problem. Turbulent motion is often described as fluid motion with increased viscosity. In this understanding, the fluctuating term is typically moved to the right side of Equation (6.5) in order to interpret it as a turbulent stress term that acts in addition to the viscous stress term. However, a different notation is chosen here in order to point out that the interference of turbulent fluctuations with the mean flow originates from the nonlinearity in the Navier–Stokes equations. The equations for the instantaneous velocity fluctuations, u_i, are given by Jovanović [63]:

$$\frac{\partial u_i}{\partial x_i} = 0 \quad , \tag{6.6}$$

$$\frac{\partial u_i}{\partial t} + \overline{U}_k \frac{\partial u_i}{\partial x_k} + u_k \frac{\partial \overline{U}_i}{\partial x_k} + \frac{\partial u_i u_k}{\partial x_k} - \frac{\partial \overline{u_i u_k}}{\partial x_k} = -\frac{1}{\rho}\frac{\partial p}{\partial x_i} + \nu \frac{\partial^2 u_i}{\partial x_k \partial x_k} \quad . \tag{6.7}$$

Manipulation of these last equations leads to the transport equation of the turbulent stresses, $\overline{u_i u_j}$ [63]:

$$\begin{aligned} \frac{\partial \overline{u_i u_j}}{\partial t} + \overline{U}_k \frac{\partial \overline{u_i u_j}}{\partial x_k} = & \underbrace{-\overline{u_i u_k}\frac{\partial \overline{U}_j}{\partial x_k} - \overline{u_j u_k}\frac{\partial \overline{U}_i}{\partial x_k}}_{P_{ij}} \underbrace{- \frac{\partial}{\partial x_k}\overline{u_i u_j u_k}}_{T_{ij}} \\ & \underbrace{-\frac{1}{\rho}\left[\overline{u_i \frac{\partial p}{\partial x_j}} + \overline{u_j \frac{\partial p}{\partial x_i}}\right]}_{\Pi_{ij}} -2\,\nu \underbrace{\overline{\frac{\partial u_i}{\partial x_k}\frac{\partial u_j}{\partial x_k}}}_{\epsilon_{ij}} + \nu \underbrace{\frac{\partial^2 \overline{u_i u_j}}{\partial x_k \partial x_k}}_{D_{ij}} \quad . \end{aligned} \tag{6.8}$$

This equation contains three unclosed terms, T_{ij}, Π_{ij} and ϵ_{ij}. A short explanation of the physical meaning of the individual terms that describe the substantial derivative of $\overline{u_i u_j}$ is as follows:

P_{ij}: production of $\overline{u_i u_j}$ by the mean flow

T_{ij}: turbulent transport

Π_{ij}: velocity–pressure gradient correlations

ϵ_{ij}: turbulent dissipation, which is the essential feature of turbulence; this can be imagined as energy transfer through a cascade process from the large scales to the smallest ones in which the energy at the final stage is dissipated into heat

D_{ij}: viscous diffusion.

Based on the statistical concept, the conservation equation of the kinetic energy of turbulent flows can be derived as shown by Hinze [54]. For this purpose, the decomposition of the variables introduced in Equation (6.3) is complemented by the following rule:

$$U_iU_i = \overline{U}_i\overline{U}_i + 2\overline{U}_iu_i + u_iu_i = \overline{U}_i\overline{U}_i + 2\overline{U}_iu_i + q^2 \quad . \tag{6.9}$$

Applying these relations to Equation (2.11) leads to

$$\underbrace{\frac{1}{2}\frac{\partial \overline{U}_i\overline{U}_i}{\partial t} + \frac{1}{2}\frac{\partial \overline{q^2}}{\partial t}}_{I} = -\underbrace{\frac{\partial}{\partial x_k}\overline{U}_k\left(\frac{\overline{P}}{\rho} + \frac{\overline{U}_i\overline{U}_i}{2}\right)}_{II} + \underbrace{\nu\frac{\partial}{\partial x_k}\left[\overline{U}_i\left(\frac{\partial \overline{U}_k}{\partial x_i} + \frac{\partial \overline{U}_i}{\partial x_k}\right)\right]}_{III} \tag{6.10}$$

$$\underbrace{-\nu\left(\frac{\partial \overline{U}_k}{\partial x_i} + \frac{\partial \overline{U}_i}{\partial x_k}\right)\frac{\partial \overline{U}_i}{\partial x_k}}_{IV} - \underbrace{\frac{\partial}{\partial x_k}\overline{u_k\left(\frac{p}{\rho} + \frac{1}{2}q^2\right)}}_{V} - \underbrace{\frac{\partial}{\partial x_k}\overline{U}_i\overline{u_ku_i}}_{VI} - \underbrace{\frac{1}{2}\frac{\partial}{\partial x_k}\overline{U}_k\overline{q^2}}_{VII} +$$

$$\underbrace{\nu\frac{\partial}{\partial x_k}\left[\overline{u_i\left(\frac{\partial u_k}{\partial x_i} + \frac{\partial u_i}{\partial x_k}\right)}\right]}_{VIII} - \underbrace{\nu\overline{\left(\frac{\partial u_k}{\partial x_i} + \frac{\partial u_i}{\partial x_k}\right)\frac{\partial u_i}{\partial x_k}}}_{IX} \quad ,$$

where $k = q^2/2$ represents the kinetic energy of turbulence.
The individual terms in Equation (6.11) can be physically interpreted as follows (where all descriptions are defined as per unit mass and time):

- I: local change of kinetic energy
- II: change in convective transport of the total energy due to the mean flow
- III: work done by the viscous stresses of the mean flow
- IV: direct dissipation, ϵ_d
- V: convective transport of the total energy due to turbulent fluctuations
- VI: work done by the turbulent stresses
- VII: work of deformation by the turbulent stresses
- VIII: work done by the viscous shear stress of the turbulent motion
- IX: dissipation by the turbulent motion, ϵ.

It is clear that the processes governing the above equation are related to both the mean flow and turbulence. The contribution of the turbulent motion will be addressed specifically in the following discussion. It can be explicitely described in terms of the transport equation of the kinetic energy of turbulence [63]:

$$\frac{\partial k}{\partial t} + \overline{U_k}\frac{\partial k}{\partial x_k} = \underbrace{-\overline{u_i u_k}\frac{\partial \overline{U_i}}{\partial x_k}}_{P_k} \underbrace{-\frac{\partial}{\partial x_k}\overline{u_k\left(\frac{u_s u_s}{2} + \frac{p}{\rho}\right)}}_{T_k} \underbrace{-\nu\overline{\frac{\partial u_i}{\partial x_k}\frac{\partial u_i}{\partial x_k}}}_{\epsilon} + \underbrace{\nu\frac{\partial^2 k}{\partial x_k \partial x_k}}_{D_k} \quad . \tag{6.11}$$

This equation can also be derived by contraction of Equation (6.8) on setting $i = j$. The the physical meaning of the individual terms are related in both equations and the nomenclature is used accordingly.

6.1.2 Basic equations for laminar flow with superimposed small disturbances

The stability of laminar flow is traditionally analysed considering an underlying laminar field that is exposed to two-dimensional perturbations applying the method of small disturbances [106]. This procedure forms the basis for the derivation of the Orr–Sommerfeld equation. In spite of simplifying assumptions in its derivation, this equation is found to determine the criteria leading to transition of boundary layer flows in a satisfactory manner. However, this ansatz fails in predicting the critial conditions of internal flows [106].
Compared with the traditional procedure, Jovanović *et al.* [65] use a more general approach and express the motion of laminar flow in a statistical frame. The resulting basic equations governing the motion of laminar flow that is exposed to disturbances are summarized as follows.
The incompressible laminar flow field with superimposed disturbances is governed by the continuity equation and the Navier–Stokes equations given in Equations (2.4) and (2.5). Starting from these equations, the method of separating the instantaneous velocity U_i and the pressure P into the mean-laminar contribution, U_i' and P', and disturbances superimposed on it, u_i' and p', is introduced:

$$U_i = U_i' + u_i', \quad P = P' + p' \quad . \tag{6.12}$$

This approach is related to that applied to turbulent flow. There, the instantaneous flow quantities are decomposed into a time-averaged part, which differs from the laminar contribution, and a fluctuating deviation. In general, the derivation of the transport equations is similar in both flow regimes. However, it differs in the fact that the disturbances in the laminar regime are assumed to be much smaller than the corresponding quantities of the underlying flow:

$$u'_i << U'_i, \quad p' << P' \quad . \tag{6.13}$$

Thus, the influence of products of fluctuating quantities is supposed to vanish. This fact ensures that the motion of the mean flow is not influenced by the disturbances and is governed by the simple continuity (2.4) and Navier–Stokes equations (2.5). Further, the equations for the disturbances are derived by applying the previous assumption [65]:

$$\frac{\partial u'_i}{\partial x_i} = 0 \quad , \tag{6.14}$$

$$\frac{\partial u'_i}{\partial t} + U'_k \frac{\partial u'_i}{\partial x_k} + u'_k \frac{\partial U'_i}{\partial x_k} = -\frac{1}{\rho}\frac{\partial p'}{\partial x_i} + \nu \frac{\partial^2 u'_i}{\partial x_k \partial x_k} \quad . \tag{6.15}$$

By systematic manipulation of Equations (6.15) and (6.14), the transport equations for the "apparent stresses" $\overline{u'_i u'_j}$ can be derived and are given by Jovanović *et al.* [65]:

$$\frac{\partial \overline{u'_i u'_j}}{\partial t} + U'_k \frac{\partial \overline{u'_i u'_j}}{\partial x_k} = \underbrace{-\overline{u'_j u'_k}\frac{\partial U'_i}{\partial x_k} - \overline{u'_i u'_k}\frac{\partial U'_j}{\partial x_k}}_{P_{ij}}$$

$$\underbrace{-\frac{1}{\rho}\left[\overline{u'_j \frac{\partial p'}{\partial x_i}} + \overline{u'_i \frac{\partial p'}{\partial x_j}}\right]}_{\Pi_{ij}} \underbrace{-2\nu \overline{\frac{\partial u'_i}{\partial x_k}\frac{\partial u'_j}{\partial x_k}}}_{\epsilon_{ij}} + \underbrace{\nu \frac{\partial^2 \overline{u'_i u'_j}}{\partial x_k \partial x_k}}_{D_{ij}} \tag{6.16}$$

Obviously, the above set of equations are related to the equations for the turbulent stresses (6.8). However, their complexity is reduced and they only imply two different types of unknown correlations: the velocity–pressure gradient correlations, Π_{ij}, and the dissipation correlations, ϵ_{ij}.

6.1.3 Interpretation of the statistical flow properties using invariant theory

Based on the statistical description of turbulent flow, Lumley and Newman [82] use kinematic considerations for the construction of a space that bounds all physically realistic states of turbulence. This space is defined in terms of the two independent scalar invariants of the anisotropy tensor, a_{ij}, and is shown to enable the transparent illustration of outstanding properties of turbulent flow.

To quantify the anisotropy in $\overline{u_i u_j}$, Lumley and Newman [82] introduce the tensor

$$a_{ij} = \frac{\overline{u_i u_j}}{\overline{q^2}} - \frac{1}{3}\delta_{ij} \quad , \tag{6.17}$$

and its scalar invariants:

$$\begin{aligned} \mathrm{II} &= a_{ij}a_{ji} \quad , \\ \mathrm{III} &= a_{ij}a_{jk}a_{ik} \quad , \end{aligned} \tag{6.18}$$

where δ_{ij} represents the Kronecker delta. With a_{ij} it is possible to distinguish the anisotropy of $\overline{u_i u_j}$ from all other flow quantities. Physically, II can be understood as the magnitude of anisotropy, III as its type. Thus, for isotropic turbulence $a_{ij} = 0$ holds and consequently II = III = 0.
A configuration of $\overline{u_i u_j}$ that permits the anisotropy to be quantified is the case of axisymmetric turbulence. In such turbulence the invariants are described by [63]

$$\mathrm{II} = \frac{3}{2}\left(\frac{4}{3}|\mathrm{III}|\right)^{2/3} \quad . \tag{6.19}$$

For the case of two-component turbulence the relationship between the invariants gives rise to [63]

$$\mathrm{II} = \frac{2}{9} + 2\,\mathrm{III} \quad . \tag{6.20}$$

In Figure 6.1 these relations are plotted. The resulting "triangle" defines the anisotropy invariant map and covers all realizable states of turbulence. The intersections of the lines representing axisymmetry and two-componentality in the stress tensor define the corners of the map. The upper branch of the map at the position (III, II) = (2/9, 2/3), where the anisotropy is maximal, is referred to as the one-componental state of turbulence. The lower left corner of the map at the position (III, II) = (−1/36, 1/6) denotes the isotropic two-componental state of turbulence.
In the present investigation, the anisotropy invariant map is used to visualize the statistical properties of wall-bounded flows. For this flow, fluctuations are forced to be two-componental when approaching the wall while they reach an almost isotropic state in the center of the domain. The trajectory for turbulent channel flow [5] is sketched in Figure 6.1 to illustrate the characteristic development. When moving away from the wall, it can be seen, that the anisotropy in the flow first increases and reaches its maximum at the edge of the viscous sublayer ($x_2^+ \simeq 8$) [63]. Further, the flow approaches the isotropic state in developing along the right branch

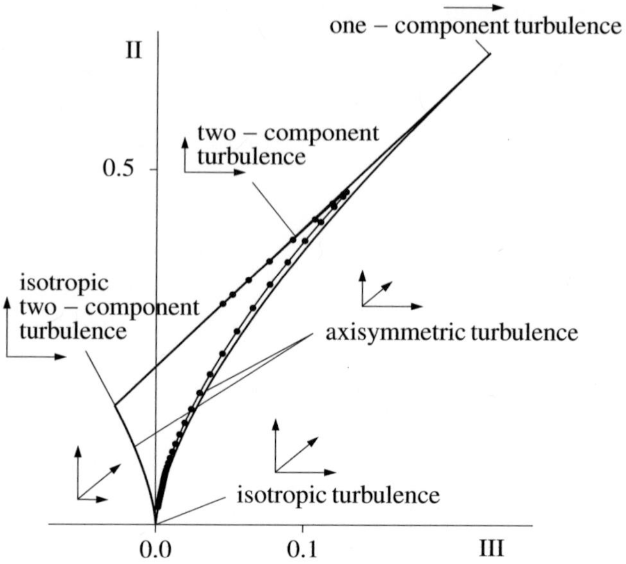

Figure 6.1: Anisotropy–invariant map [82]. The boundaries and corner points of the map represent outstanding states of turbulence that are assigned in the figure. Additionally, the arrows illustrate possible configurations of the stress tensor represented by the boundaries of the map. The trajectory for turbulent channel flow at the friction Reynolds number $Re_\tau = 180$ [5] (line with dots) shows a characteristic development of wall-bounded flows.

of the map indicating an almost axisymmetrical configuration of the stresses.
The development of Lumley and Newman originally aims at a description of turbulence. So far the findings have strong and long lasting implications for the development of turbulent closure approximations that have the potential to describe a wide range of shear flows [63]. It was also shown that the invariant map represents a powerful tool for the illustration of mechanisms that accompany drag reduction in turbulent flows. The success of various turbulent control strategies for wall-bounded flows was related to the level of anisotropy in the turbulent stresses that is achieved at the wall [41].
The implication of high anisotropy and the axisymmetric state of the stresses on the energy dissipation in the flow is discussed next. In the present flow configurations, x_1 is always defined as the direction of the mean flow and axisymmetry in the stresses is limited to invariance of the stresses under rotation about this axis. In

the case of full axisymmetry in such a flow, the stress tensor can be expressed as follows [63]:

$$\overline{u_i u_j} = F\delta_{ij} + G\lambda_i\lambda_j \quad , \qquad \lambda_i\,(1,0,0) \quad , \tag{6.21}$$

with F and G being scalar functions.
Next to turbulent flow, the anisotropy map can be also used to analyse statistically described laminar flow. In this way, Jovanović *et al.* [65] derive constraints for the persistance of stable laminar flow and consequently for the intended delay of laminar to turbulent transition.
Thus, the representation of the stresses in the anisotropy–invariant map is expected to be a powerful tool for the aim of investigating mechanisms leading to low energy dissipation in internal flows. In particular, it enables the simultaneous investigation of the configuration of the stresses in the different flow regimes.

6.2 Physical models for reduced friction losses of wall-bounded flows

The theoretical considerations followed within this chapter are based on the statistical properties of disturbances in laminar flow and of fluctuations in fully developed turbulent flow. In this framework, analytical constraints can be formulated that lead to reduced friction losses. In general, the implementation of these constraints is supposed to be associated with benefits in internal and external flows, while the former flow situation is the focus of this work.

6.2.1 Kinematic consideration of velocity fluctuations

The analysis is based on the theoretical findings from Jovanović and Hillerbrand [67] for the peculiar properties of velocity fluctuations in the near-wall region. In their investigation, the implication of axisymmetry in the fluctuations is studied using kinematic considerations.
For this purpose, the flow above a plane wall is considered were x_1 is the direction of the mean flow, x_2 is the wall-normal coordinate with origin on the wall and x_3 is the spanwise coordinate. A Taylor series expansion of the instantaneous velocity fluctuations reads as [87]:

$$\begin{aligned} u_1 &= a_1 x_2 + a_2 x_2^2 + ... \\ u_2 &= b_1 x_2 + b_2 x_2^2 + ... \\ u_3 &= c_1 x_2 + c_2 x_2^2 + ... \end{aligned} \tag{6.22}$$

where the coefficients a_i, b_i and c_i are functions of time and the space coordinates x_1 and x_3. In order to satisfy the continuity equation close to the wall, the coefficient b_1 vanishes, $b_1 \rightarrow 0$.
For axisymmetric disturbances that are invariant to rotation about the streamwise x_1 coordinate, the following relation is obtained for the statistics of the velocity derivatives of nth order [45]:

$$\overline{\left(\frac{\partial^n u_1}{\partial x_2^n}\right)^2} = \overline{\left(\frac{\partial^n u_1}{\partial x_3^n}\right)^2}, \quad \overline{\left(\frac{\partial^n u_2}{\partial x_2^n}\right)^2} = \overline{\left(\frac{\partial^n u_3}{\partial x_3^n}\right)^2}, \quad \overline{\left(\frac{\partial^n u_2}{\partial x_3^n}\right)^2} = \overline{\left(\frac{\partial^n u_3}{\partial x_2^n}\right)^2}, \tag{6.23}$$

On inserting the series expansion (6.22) into these expressions, and comparing terms corresponding to the same power of x_2, it follows that all coefficients a_i, b_i and c_i must vanish in order to satisfy constraints imposed by axisymmetry. Thus, it is concluded, that velocity fluctuations, and therefore the turbulent properties of the flow, must vanish in the close proximity of the solid boundary, for the assumed state of the fluctuations [67]. This finding also implies that the dissipation due to turbulent fluctuations vanishes at the wall for this particular state of the stresses.
In general, the stresses at the wall are forced to be two-component due to constraints imposed by the continuity equation. The additional constraint of axisymmetry leads to the one-component state. In the anisotropy invariant representation of the stresses, the one-component limit denotes the state of maximum anisotropy, since II denotes the magnitude of anisotropy (see Figure 6.1). Using data from direct numerical simulations of turbulent wall-bounded flows, Jovanović and Hillerbrand [67] correlate the dissipation at the wall with the magnitude of anisotropy. In Figure 6.2 their original illustration is replotted and supplemented with additional data. The extrapolated trend in the data (dashed line) supports the conclusions from the theoretical findings: If axisymmetry in the stresses at the wall, and thus the one-component limit, is reached, the dissipation due to turbulent fluctuations vanishes.
From the findings of Jovanović and Hillerbrand [67], strategies for reduced energy dissipation in the different flow regimes can be derived. Their considerations focus on the statistical properties of turbulent flow. However, the similarities in the statistical description results in a related statement for the laminar regime: disturbances cannot be amplified if they are axisymmetrical. The physical models for energy savings in both flow regimes are formulated in the following paragraphs.

6.2.2 Delay of transition to turbulence

In Figure 4.2 it was demonstrated that the flow resistance due to friction forces in a cylindrical pipe is much lower in laminar flow than in the turbulent regime.

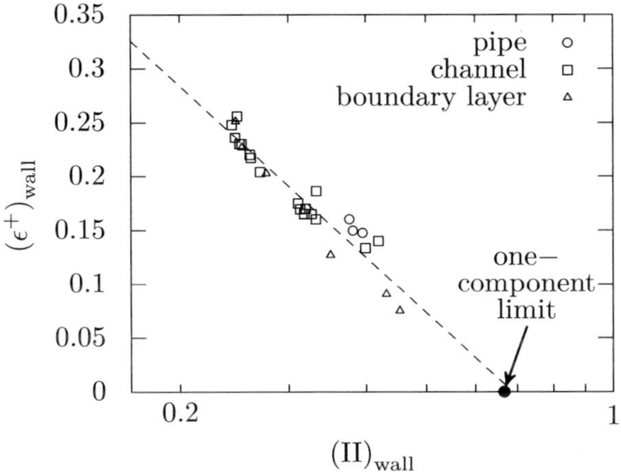

Figure 6.2: Turbulent dissipation rate at the wall *versus* the magnitude of anisotropy at the wall following the illustration of Jovanović and Hillerbrand [67]. Numerical data from wall-bounded flows are used to extrapolate the trend. Pipe flow: Eggels *et al.* [32], Unger and Friedrich [112], Fukagata and Kasagi [42]. Channel flow: Kim *et al.* [69], Moser *et al.* [88], Alamo and Jimenez [5], Gilbert and Kleiser [47], Horiuti [57], Iwamoto *et al.* [59]. Flat plate boundary layer: Spalart [109], Spalart [108]. The dissipation rate is normalized with the wall shear velocity and the kinematic viscosity of the fluid.

Additionally, this difference increases with increasing Reynolds number. Thus, it is logical to conclude that keeping flow laminar in situations in which it will usually turn into turbulence is very promising if energy losses are to be minimized.
In Section 6.1.2 the statistical discription of laminar flow subjected to disturbances was introduced. The presence and appearance of such disturbances form the basis for considerations of the laminar to turbulent transition process in different types of wall-bounded flows [65, 67]. In the previous paragraph it was shown that the dissipation term in Equation (6.16) vanishes in the near-wall region, if the velocity fluctuations are forced to be axisymmetrical. Thus, it can be concluded that laminar flow in the near-wall region is stable to any level of disturbances if they appear in the statistically axisymmetric state. The same conclusion is also drawn by Jovanović *et al.* [65] in a theoretical investigation on the persistence of laminar flow in the flat plate boundary layer. However, it is found that reaching the one-component state at the wall is only a necessary but not sufficient condition for the disturbances, if the stability of the flow is in question. In their formulation of the transition criterion,

they assume laminar flow to be exposed to small axisymmetrical disturbances that are invariant under rotation about the axis of the main flow direction. Further, they consider local equilibrium to exist between the production and dissipation. From the dissipation equation, by requesting that ϵ must always be positive, they derive a criterion that determines conditions that guarantee maintenance of the laminar regime. The limit for the ensurance of laminar flow is quantified in terms of the magnitude of the anisotropy in the disturbances in the free stream, II_∞ [65]:

$$II_\infty \geq 0.141 \quad . \tag{6.24}$$

In general, this formulation is supposed to be valid for a wide range of flow situations, if the superimposed disturbances are axisymmetrical. The authors tested the validity of the criterion for transition and the breakdown to turbulence in the flat plate boundary layer. Comparison of the theoretically derived criterion with experimental data shows good agreement. Experimental findings and recent results from numerical investigations of transition in the flow through a cylindrical pipe also support these theoretical results: the laminar to turbulent transition process in the flow is characterized by decreasing anisotropy in the disturbances [33, 93].
These findings are used to derive a physical model of how the laminar flow regime can persist at high Reynolds numbers, which is summarized in Figure 6.3. If the disturbances superimposed on laminar flow are axisymmetrical and reach a high level of anisotropy, they cannot be amplified and the transition to turbulence is delayed. In principal, the described strategy is supposed to hold for external and internal flows. The aim is to achieve the axisymmetrical configuration of the disturbances, which leads to reduced friction losses, due to geometrical modification in the fluid–solid boundary. This approach is considered to result in practically realizable solutions. For external flows, an appropriately designed surface morphology is expected to result in the target state (left branch of Figure 6.3), while it might be provoked by the cross section shape of ducts in internal flows (right branch of Figure 6.3). The latter aspect is addressed in numerical investigations in the following chapters. Since the flow is controlled passively, reduced friction losses will directly result in energy savings, as discussed in Section 4.4.

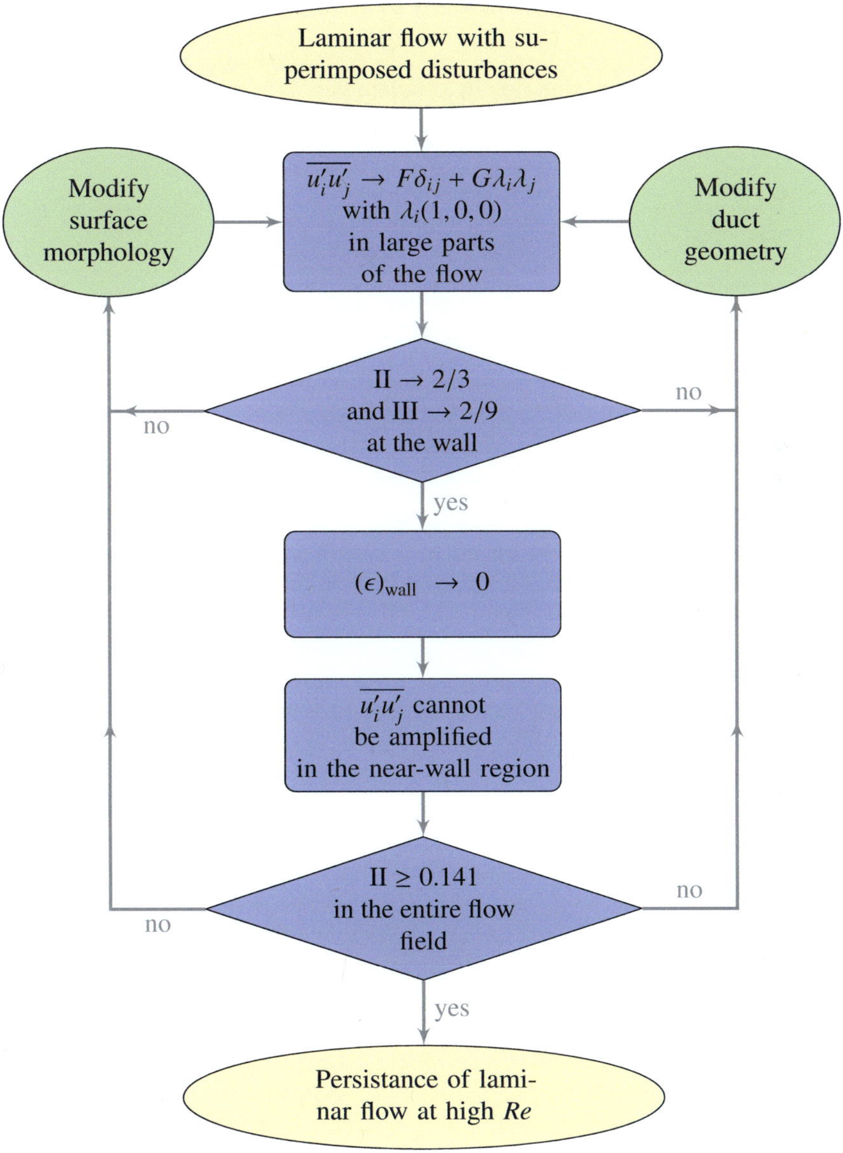

Figure 6.3: Physical model aiming at low friction losses and persistence of the laminar regime in wall-bounded flows.

6.2.3 Reduction of dissipation in turbulent flows

Laminar flow with superimposed disturbances and turbulent flow have common properties. However, the situation in turbulent flow is more complex: the fluctuations modify the mean flow, as can be seen in Equation (6.5).
The entire energy dissipation rate in turbulent flow, Φ, gives rise, according to the corresponding terms in Equation (6.11), to

$$\Phi = \underbrace{\nu\left(\frac{\partial\overline{U_i}}{\partial x_j} + \frac{\partial\overline{U_j}}{\partial x_i}\right)\frac{\partial\overline{U_i}}{\partial x_j}}_{\epsilon_d} + \underbrace{\overline{\nu\left(\frac{\partial u_i}{\partial x_j} + \frac{\partial u_j}{\partial x_i}\right)\frac{\partial u_i}{\partial x_j}}}_{\epsilon} \quad . \tag{6.25}$$

The contribution controlled by the mean flow is referred to as direct dissipation, ϵ_d, and the contribution originating from the fluctuating motion is termed as turbulent dissipation, ϵ. The order of magnitude analysis shows that direct dissipation, ϵ_d, is of the order $O\left(U^2/L^2\right)$ where L is defined by the size of the flow domain and is constant. The turbulent contribution, ϵ, is of the order $O\left(q^2/\lambda^2\right)$ and λ is the Taylor micro-scale, which decreases with increasing Reynolds number. These relationships suggest that the contribution of turbulence to Φ prevails at large Reynolds numbers. Data from direct numerical simulation of plane channel flow support this statement [38, 76]. Thus, at high Reynolds numbers, which are of particular importance in practical applications, the decrease and possible minimization of ϵ is promising for the achievement of significant reduction in the entire dissipation, Φ. The control strategy emphasized in subsequent paragraphs is derived from this deduction.
Kinematic considerations of velocity fluctuations in wall-bounded flows show that the turbulent dissipation at the wall vanishes, if the velocity fluctuations are forced to be axisymmetrical around the axis of the mean flow (see Section 6.2.1). This state of the fluctuations imply that they reach the one-component state at the wall. Frohnapfel *et al.* [41] study the impact of axisymmetric disturbances in the near-wall region on the turbulent dissipation in the entire flow field. For this purpose, numerical experiments in turbulent plane channel flow are performed. In the simulations, the flow in the vicinity of the wall is forced to be axisymmetrical by damping of the spanwise velocity fluctuations in the near-wall region. It is found, that this state of fluctuations at the wall results in a decrease in the turbulent dissipation in the entire flow field. In the simulations, the number of points in the wall-normal direction, for which the forcing is applied, is varied. A comparison of the results shows, that the decrease in the turbulent dissipation is more pronounced if the axisymmetric state extends further into the flow field.

The observations made in the numerical experiment can be summarized as follows: the constraint of axisymmetry ($\overline{u_2^2} = \overline{u_3^2}$ and $\overline{u_i u_j} = 0$ for $i \neq j$) leads to the conclusion, that the production of turbulent kinetic energy vanishes, $P_k \to 0$ (since $P_k = -\overline{u_1 u_2}\,\partial \overline{U_1}/\partial x_2$). The equilibrium consideration suggests that production and dissipation are balanced in the flow, $P_k \simeq \epsilon$, resulting in the same behavior for ϵ as for P_k, namely $\epsilon \to 0$.
The impact of changes in the turbulent dissipation rate on the flow can be discussed as follows. A general property of turbulent flow is the separation of scales. In internal flows, the largest spatial scale, L, is fixed due the restriction of the flow domain while the smallest scales are defined in terms of ϵ and ν and are in the order of the Kolmogorov length scale, η_K. The strength of turbulence can be expressed in terms of the spectral separation in the flow, L/η_K, and might be defined as function of the turbulent Reynolds number, R_λ. Vanishing turbulence is accompanied by vanishing spectral separation, $L/\eta_K \to 1$ as $R_\lambda \to 0$, and strong turbulent activity is characterized by large spectral separation, $L/\eta_K \gg 1$ when $R_\lambda \gg 1$. The Kolmogorov length scale is connected to ϵ as follows [63]:

$$\eta_K = \left(\frac{\nu^3}{\epsilon}\right)^{1/4} \quad . \tag{6.26}$$

Obviously η_K increases as ϵ is decreased leading to a decrease of the spectral separation in the flow. Therefore we may expect that R_λ decreases with decreasing ϵ, which additionally corresponds to a reduced turbulent contribution to the total dissipation. This qualitative behavior of changes induced by decreasing ϵ leads to the conclusion that any changes induced in turbulence with the intention of decreasing ϵ will have a strong tendency to provoke laminarization in the flow.
Similar conclusions can be drawn for fully developed flows, if the transport equation of the mean flow (6.5) is considered and the turbulent fluctuations interacting with the mean flow field are assumed to be axisymmetrical. In this state, the stresses can be expressed in the linear fashion given in Equation (6.21). For such turbulence, Equation (6.5) can be rewritten as follows [78]:

$$\frac{\partial \overline{U}_i}{\partial t} + \overline{U}_k \frac{\partial \overline{U}_i}{\partial x_k} = -\frac{1}{\rho}\frac{\partial}{\partial x_i}\underbrace{\left(\overline{P} + \frac{1}{3}\rho\overline{q^2} + \frac{2}{3}\rho B\right)}_{P^*} + \nu \frac{\partial^2 \overline{U}_i}{\partial x_k \partial x_k} \quad . \tag{6.27}$$

In this equation, turbulent fluctuations have a similar impact on the flow as the pressure. Both quantities can be expressed together as a modified pressure term,

P^*. For fully developed flow in channels, pipes and ducts, this equation leads to solutions that coincide with solutions for laminar flow.

The tendency towards laminarization for producing reduced friction losses is supported by many flow situations when looking at the evolution of turbulence across the anisotropy invariant map. An example in this respect is the drag reduction due to the addition of dilute polymers [68]. The same can also be concluded from numerous examples of wall-bounded flows, where the same or similar drag reducing effects are observed [38]. In all of these documented examples it turns out that achieving drag reduction and a tendency towards laminarization of the flow is accompanied by the restructuring of turbulence towards the statistically axisymmetric state with invariance to rotation about the axis aligned with the main flow.

This analysis of the effect of axisymmetry in the stresses on turbulent flow is summarized schematically in Figure 6.4 and forms the basis of the physical model for reduced friction losses in this flow regime. The major goal is to approach the axisymmetric state of the stresses, which implies that the fluctuations at the wall tend towards the one-component state. As a consequence, the turbulent dissipation and also spectral separation are decreased. For full axisymmetry, turbulent activity is totally suppressed, resulting in pure laminar flow and significant energy savings. The open question is how to actually achieve axisymmetry in wall-bounded flows. For this purpose, geometrical modification in the bounding walls are proposed. The present concept for turbulent flow is essentially similar to that introduced for laminar flow in Figure 6.3. Again the possibility of controlling external and internal flow is considered, which is represented by the left and right branch of the model, respectively. The following numerical studies address the impact of geometrical modification on internal flows but similar geometrical designs can be imagined to result in the required configuration of the fluctuations in both flow situations.

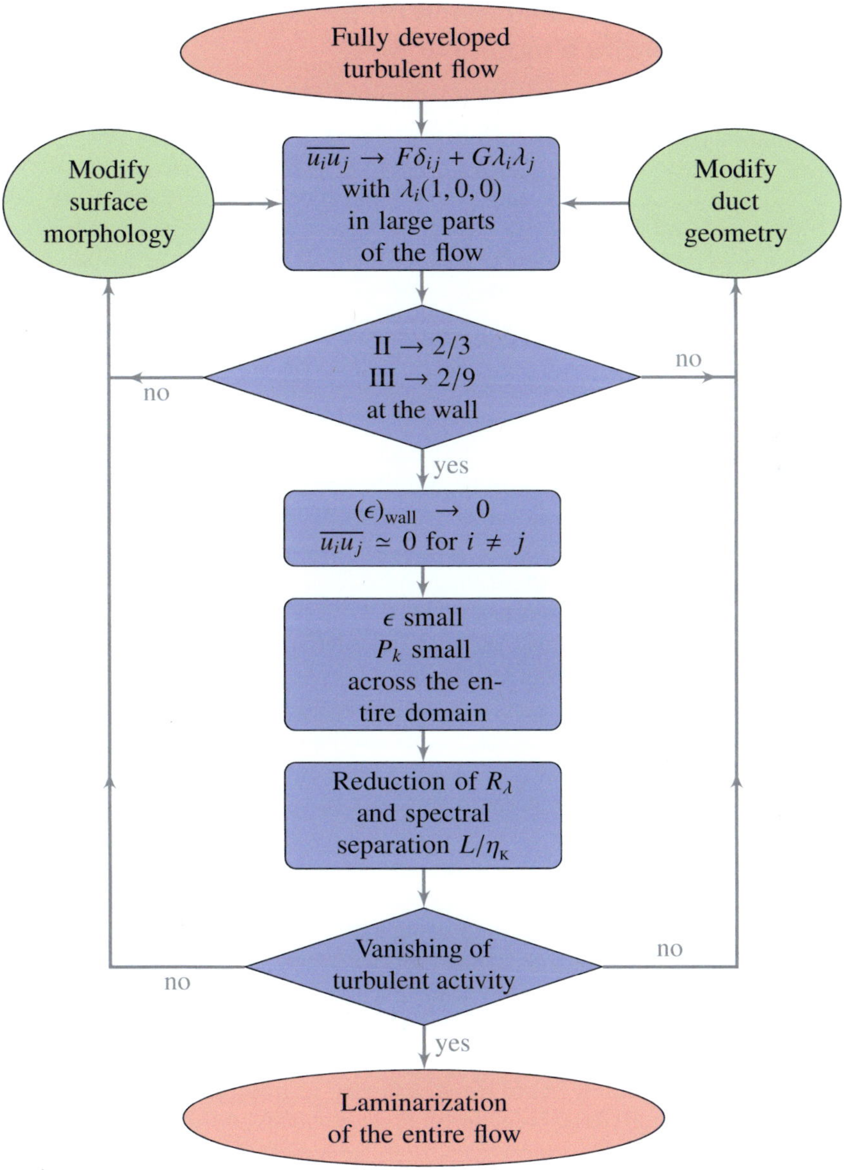

Figure 6.4: Physical model aiming at the reduction of energy dissipation and laminarization of turbulent flow.

6.2.4 Review of investigations on friction drag reduction due to surface modifications

The impact of the surface design on the friction losses of flows at higher Reynolds numbers has been studied frequently in the past. The majority of the proposed surface structures are aligned with the mean flow direction, exceptions are represented by, for example, dimples [80] or distributed roughness elements [101]. In Figures 6.3 and 6.4, axisymmetry of the stresses about the axis aligned with the main flow direction is shown to be associated with reduced energy dissipation. Geometrical modifications in the fluid–solid boundary in the form of structures that are aligned with the bulk flow and control the development of the wall normal and spanwise component of the stress tensor are proposed to provoke the desired state. Major findings for this class of surface structures in the literature are summarized as follows.

One of the earliest suggestions in this respect was presented in a patent by Kramer from 1937 [71]. He proposes thin wires to be arranged a small distance above a flat surface. The wires are parallel to each other and aligned in the flow direction. In spite of the surface area increase, an overall drag reduction is reported [38].

Since the pioneering work of Kramer, different types of structure geometries have been investigated, which are referred to as riblets. Generally, riblets are beneficially applied to turbulent external and internal flows. However, in most of these investigations the riblets are implemented in channel walls. This set-up was discussed in the context of laminar flows in Section 5.3. During corresponding numerical and experimental studies, riblets are found to reduce the friction losses of turbulent flow by up to 10% [11, 21]. It is observed that the maximum reduction is reached for a narrow range of dimensionless riblet spacings, namely for $s^+ \approx 15$. For the evaluation of s^+, the spacing s between two riblet tips is normalized with the characteristic viscous length scale calculated according to Equation (2.10), yielding $s^+ = s\,u_\tau/\nu$. For larger riblet spacings, drag increase is observed. Note that this dependence on s^+ differs from the findings in laminar flow, where drag reduction can only be observed if the riblet structure is wide enough.

In the literature, different mechanisms are discussed to lead to the observed friction behavior of turbulent flow over riblets. However, they actually agree, that the flow is beneficially influenced in the near-wall region within valleys between the structures [21, 37]. The effect of riblets can also be interpreted in the context of the strategy proposed in the previous section, as shown by Frohnapfel *et al.* [41]. Similar conclusions can be drawn from recent numerical results from Fink [37] for channel flow over the riblet geometry shown in Figure 6.5. In the direct numerical simulations, the friction Reynolds number, $Re_\tau = u_\tau\delta/\nu$, which is based on the friction velocity, u_τ, and half of the channel height, δ, is set to $Re_\tau \simeq 180$. The numerical work accompanies experiments performed by Güttler [50]. For $s^+ = 17$, the experimental as well as the numerical investigation result in $DR \simeq 6\%$, while for $s^+ = 30$ drag increase is observed ($DR \simeq -3\%$). The observed decrease in the friction loss is supposed to be associated with the expected trend towards axisymmetry in the stresses. In Figure 6.5 the data along the symmetry line between two riblets are plotted in the anisotropy invariant map. For comparison, the wall point arising from flat channel flow at a similar Reynolds number [5] is marked. In fact, for $s^+ = 17$, a strong tendency to reach axisymmetry in the near-wall region and thus the one-component limit at the wall is observed. In the drag increasing case, this trend is also present when comparison is made to the flow over flat channel walls (see black arrow in Figure 6.5). However, it is significantly weaker and obviously the resulting benefits do not overcome the increase in the wetted area and possible negative effects on the riblet tip.

The impact of riblets that are implemented in cylindrical pipes is also investigated by several authors. The drag reduction obtained is found to be similar to that in channel flow, but appears at slightly higher values of s^+ [81, 89].

From the results obtained for turbulent flow over riblets, it can be concluded that the impact of these structures is restricted since the maximal drag reduction is comparatively low. The laminarization of turbulent flow is not observed.

Along with the strategies for reduced energy dissipation presented in Section 6.2, a surface topology consisting of square grooves was developed. The size of the grooves is significantly smaller than that of riblets, namely in the order of 5 viscous length scales. Analysis of the impact of this type of surface morphology on turbulent channel flow is presented by Frohnapfel [38]. The investigation of the flow field within surface embedded grooves arising from DNS indeed shows a strong tendency of the flow to reach the statistical axisymmetric state in the near-wall region. However, the beneficial effect appearing within the grooves does not persist along the entire wall.

The turbulent flow through ducts with embedded corner regions along the perimeter is investigated numerically by Lammers et al. [78]. The spacing of the resulting structures is significantly larger than that of the riblets of grooves. Changes in the flow field that are linked with the modified cross section shape are discussed in the statistical framework proposed within this chapter. These findings are closely related to the present investigation and will be discussed further in Chapter 9.

Compared with the summarized results for turbulent flow, investigations concerning the impact of surface structures on the stability of laminar flow are only rarely presented in the literature and are mainly conducted experimentally. In this context, riblets are tested for their potential to delay the laminar to turbulent transition of boundary layer flows [48, 77] and of fully developed pipe flows [77]. However, the

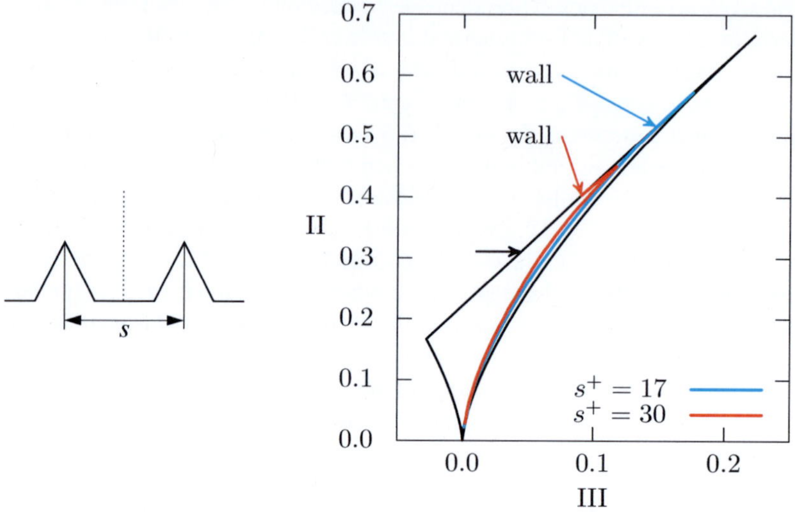

Figure 6.5: Left: Sketch of the considered riblet geometry. Right: Anisotropy invariant mapping of the stresses in turbulent channel flow over riblets at $Re_\tau \simeq 180$. DNS results for a drag reducing ($s^+ = 17$) and a drag increasing ($s^+ = 30$) configuration are shown [37]. The trajectories are plotted along the symmetry line between two riblets which is marked by a dotted line in the sketch. The wall point for flat channel flow at $Re_\tau \simeq 180$ [5] whose trajectory is plotted in Figure 6.1 is highlighted by a black arrow for comparison.

results provide no clear answer about changes in the transition Reynolds number and in the physical properties of the flow due to the presence of the riblets. In another experimental investigation, surface embedded grooves show indications of stabilizing laminar channel flow [64].

6.3 Conclusions for further investigation of internal flows

In Section 6.2, flow states that are capable of provoking flow laminarization and delay of transition, and consequently resulting in low energy dissipation, were presented. A common mechanism was found to be responsible for benefits in both flow regimes. The next task is to investigate whether geometrical properties of the flow domain can lead to the described behavior, as is proposed in Figures 6.4 and 6.3.

In general, the desired state of axisymmetry in the stresses is more likely to be achieved in duct flows than in channel flows: the cross section shape of the duct directly influences the flow in the entire domain while in plane channel flow it is only manipulated in the near-wall region due to structured walls. The invariant representation of the turbulent stresses in Figure 6.6 supports the arguments about differences between channel and duct flows. Obviously, the magnitude of anisotropy (II) and the axisymmetry at the wall is increased in circular pipe flow compared with channel flow. In addition, the details of the trajectories along the right boundary of the invariant map (Figure 6.6 right) show a stronger tendency towards the axisymmetric state for the circular pipe flow. The geometrical configuration of the circular pipe provokes the stress tensor $\overline{u_i u_j}$ to follow axisymmetry by definition at the centerline of the pipe. The observed impact of geometrical properties of duct flows is in agreement with findings from Pfenniger [96]: in experimental investigations of a pipe flow he observes laminar flow to persist up to very high Reynolds numbers, for which channel flow typically turns to turbulence. Based on these arguments, the subsequent chapters will focus on the investigation of duct flows.

Besides these physical reasons, the consideration of duct flows also addresses the general formulation of the optimization problem in Section 4.4, namely the minimization of the dissipation arising from a fixed flow rate. The difference from the findings for pure laminar flow may be highlighted in this respect. There, the cylindrical pipe is optimal [107] and benefits can be obtained only for the special case of channel flow.

In order to achieve reduced energy dissipation in pipe flows, the surface of the pipe has to be designed to achieve an extension of the axisymmetry in $\overline{u_i u_j}$ from the core region towards the wall of the pipe. For this purpose, structures whose scales are significantly increased compared with the previously mentioned riblets are proposed. Initial attempts towards this goal are made by Lammers *et al.* [78] in the scope of the Erlangen pipe concept. His findings confirm the arguments about the strong impact of geometrical properties in the flow domain on the axisymmetrical behavior of the stresses.

Within the present investigation, direct numerical simulations are performed for the investigation of the flow through ducts of different cross section shapes. This approach allows a detailed study of the properties in the Reynolds stress field arising from the cross section configuration and enables a comparison with the theoretically derived mechanisms leading to reduced energy dissipation.

Based on the findings in the previous sections, a plan for the numerical procedure consisting of 5 steps is elaborated and summarized in Figure 6.7. The goal of the present study is, when starting from the laminar and turbulent flow states in

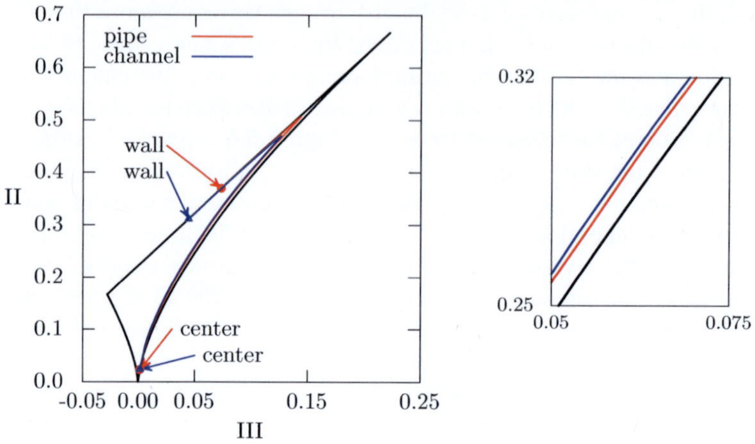

Figure 6.6: Anisotropy invariant mapping for the stresses arising from DNS of turbulent pipe and channel flow for the same friction Reynolds number $Re_\tau = 180$. Pipe flow: Eggels *et al.* [32]. Channel flow: Alamo and Jimenez [5].

step **(1)** to develop duct shapes that stabilize the flow leading to the persistence of laminar flow or the laminarization of turbulent flow **(5)**. This strategy is motivated by the fact that a common mechanism is theoretically shown to lead to these goals. However, the numerical prediction of flow laminarization and transition is demanding and not straightforward.

In previous analyses of transition in circular pipe flow, it is shown that the decay of turbulent spots is a random event and the distribution of their lifetimes follows an exponential law [8, 34]. In subsequent studies, Avila *et al.* [8] perform at least 100 simulations for a certain Reynolds number starting from different initial conditions in order to clearly identify this association for the lifetimes. Based on these studies, a beneficial modification of the duct shape is supposed to lead to shorter lifetimes of turbulent spots at a certain Reynolds number. In order to test whether this behavior is achieved with a certain duct shape, similar analyses to those shown by Avila *et al.* have to be performed. This procedure requires a huge numerical effort that is outside the scope of the present study.

Laminarization of turbulent flow is studied numerically by Uhlmann *et al.* [111]. Their proposed value for the critial Reynolds number in square duct flow also involves an extensive computational effort and requires specific codes which, however, are limited to certain geometries.

The previously described theoretical investigations, which are represented by step **(2)** in Figure 6.7, form the basis for a different approach to the subject. From comparison of Figures 6.3 and Figure 6.4 is can be concluded that the same statistical property, namely axisymmetry in the stresses, leads to benefits in transitional and turbulent flow.

In order to use this similarity for the development of beneficial duct shapes, it has to be checked whether the theoretically proposed state can be provoked in both flow regimes by a certain geometrical property. For this purpose, the flow field in ducts, which are known to lead to coexisting laminar and turbulent flow regions, are investigated in step **(3)**.

Further, the impact of different geometrical properties of ducts are investigated in fully turbulent flow **(4)**. Based on the verification of the theoretically derived behavior in **(3)**, the results for turbulent flow also allow conclusions to be made for the influence of certain geometrical designs on the flow in the laminar regimes. The tendency in the flow for laminarization and the delay of transition can be discussed in this context. This approach represents a significant reduction of the numerical effort compared with the alternative direct investigation of the processes.

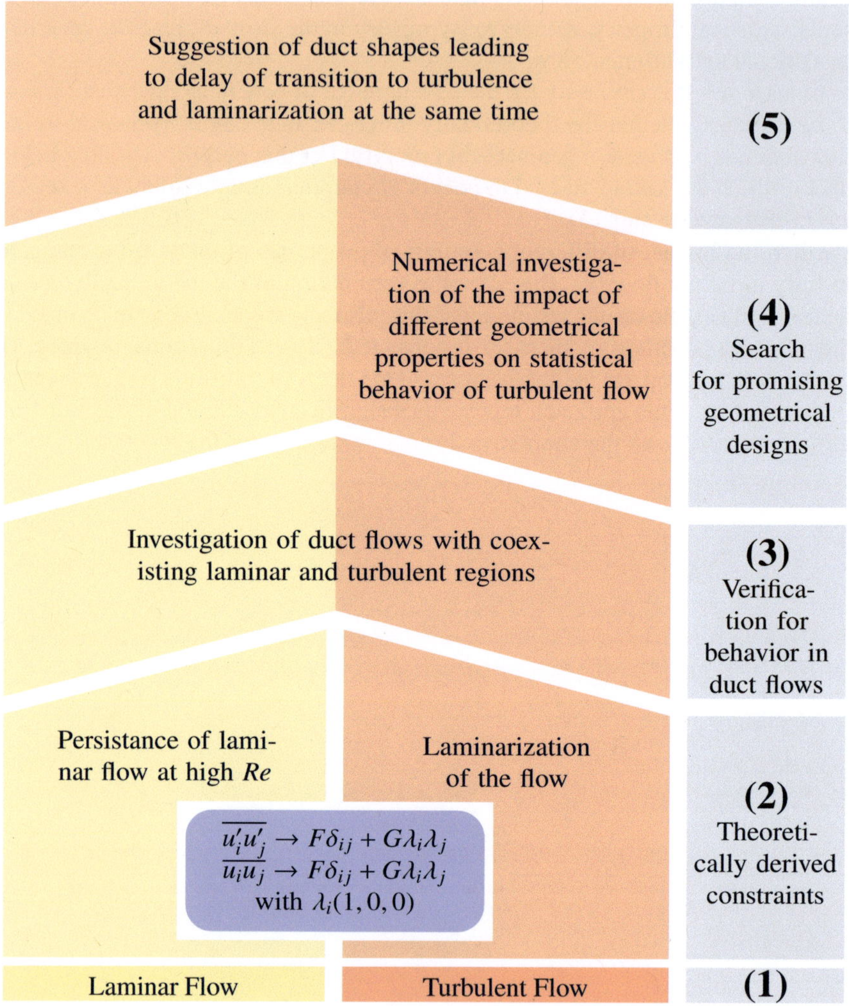

Figure 6.7: Procedure for the investigation of low energy dissipation in duct flows based on theoretical analysis for laminar and turbulent flow.

7 Presentation and evaluation of the numerical method

Within the present study, direct numerical simulations (DNS) of turbulent flow in straight non-circular ducts with different cross section shapes are carried out. It is focused to investigate the distribution of the Reynolds stress field based on these simulations. For this purpose a finite volume method in combination with polyhedral meshes is used. Since this approach is rather unusual for DNS studies, evaluation of its accuracy restrictions was considered to be essential.
Firstly, the general properties of the chosen approach together with its strong points and limits are briefly summarized and the basic set-up of the computational models together with the boundary and initial conditions are introduced.
In order to evaluate the accuracy of the method for the simulation of turbulent flow in non-circular ducts, a calculation of the flow in a square duct is performed. Owing to previous numerical investigations, detailed data for comparison are available for this duct shape. In this context, the impact of the duct length is also studied, the importance of which for the DNS of non-circular duct flows is not clearly determined in literature. The effects of the mesh type, which are of crucial importance in the context of this work, are also investigated for the square duct.
Finally, the ability of the present approach to predict flow laminarization in corner regions is tested. As mentioned previously, the investigation of coexisting laminar and turbulent flow fields forms the basis for the present procedure, which aims to develop duct shapes leading to low energy dissipation. However, since these flows are rarely studied in the literature, the ability to make quantitative comparisons is limited.

7.1 Numerical procedure for the calculation of turbulent flow in non-circular ducts

The theoretical considerations in Chapter 6 proposed that statistical axisymmetry of the stress tensor leads to a reduced energy dissipation of transitional and turbulent flows. The aim of this investigation, within a numerical framework, is

to establish whether a preferential state can be reached in turbulent flows due to certain geometrical properties of duct shapes. Thus, the numerical approach needs to be capable of accurately predicting second order turbulent statistics, namely the Reynolds stress tensor, in order to produce meaningful results. For this reason, DNS studies are performed, since this approach is not biased by errors resulting from modeling of turbulent motion. However, to perform a reliable DNS of turbulent flow, the numerical procedure has to account of the outstanding physical properties of turbulence.
Since turbulence is a three-dimensional and instationary phenomenon, its direct simulation in principle requires the resolution of all temporal and spatial scales of the flow. Since in turbulence the spectral separation is large, a considerable numerical effort will be required. For statistically fully developed internal flows that are simulated using periodic boundary conditions, the streamwise extension of the domain is critical for the proper representation of the large scales of the flow: typically two-point correlations of the solution are monitored in this context, which need to decay in the middle of the streamwise extension of the computational domain [86]. In addition, the flow in this domain has to be resolved temporally and spatially in order to capture the smallest scales of turbulence defined by the Kolmogorov time scale, τ_{K}, and length scale, η_{K}. In practice, it was observed that the smallest resolved length scale is only required to be of $\mathcal{O}(\eta_{\mathrm{K}})$ [86]. At the same time, the flow field has to be averaged temporally in order to allow statistical analysis. In this context, an averaging period, which is much larger than the time periods of the fluctuating motion, is required.

7.1.1 Numerical method

Basically, different numerical methods are applicable for the DNS of turbulent flow. These methods are shown to lead to similar results if the high resolution requirements are fulfilled [15]. One of the first attempts at the DNS of wall-bounded flows was performed by Grötzbach [49] using a finite difference method. In general, spectral methods are the most popular for the DNS of turbulent flow fields due to their high accuracy at a sufficient resolution. If all scales of the turbulent motion are resolved, the approximation error decreases exponentially with the number of grid points, N, and thus much faster than it can be observed for *e.g.* finite-volume and finite-difference methods [15]. Results from spectral simulations essentially supported the fundamental understanding of turbulence. However, the application of spectral methods is restricted to rather simple geometries, *e.g.* plane channel flow and in most cases is not applicable for engineering purposes [86]. Thus, the use of a different method is required for the simulation of more complex flow situations.

Within the present work, a second-order finite volume method is applied for the spatial approximation of the continuity (2.4) and the Navier–Stokes equations (2.5). Time discretization is achieved using the second-order implicit Euler-backward scheme. The corresponding flow solver is integrated in the open source code OpenFOAM®. Since this work concentrates on the investigation of physical mechanisms rather than on the development of a numerical code, the major properties of the numerical procedure are only briefly summarized. A more detailed description may be found in related literature [61, 113].
In the present approach, variables on the cell faces are approximated by second-order linear interpolation. The continuity and Navier–Stokes equations are coupled for the calculation of the pressure using the PISO-algorithm. This method ensures the conservation of mass and momentum by definition. However, the collocated-mesh arrangement does not ensure the conservation of the kinetic energy [36]. Nevertheless, it is commonly used for turbulence simulations in complex geometries due to its simpler form in curvilinear coordinates. A similar numerical method is introduced by Felten and Lund [35] and tested in terms of conservation errors in a large eddy simulation of a plane channel flow. They observe good agreement with reference data provided that the simulation is run at a sufficiently high mesh resolution.

7.1.2 Spatial and temporal discretization

The majority of the duct cross sections that will be discussed are discretized using meshes consisting of prism-layers along the walls and polyhedral core cells. This approach allows higher flexibility in the resolution of critical regions of complex duct shapes compared with the alternative usage of block-structured, hexahedral meshes. This behavior is nicely demonstrated by Perić [94] for the simulation of turbulent flow over riblet mounted surfaces. The meshes are initially set up to ensure $\left(\Delta x_2^+\right)_{\text{wall}} \times \left(\Delta x_3^+\right)_{\text{wall}} \leq 1.6 \times 4.6$ where x_2 is the wall-normal and x_3 is the spanwise direction. Here and subsequently, the superscript "+" stands for wall units: $l^+ = l/L_c = l\,u_\tau/\nu$. The viscous length is based on the friction velocity u_τ estimated *a priori* by the Blasius correlation [106]. This spacing ensures the resolution of the viscous sub-layer (up to $x_2^+ = 5$) with three volume cells [32, 49]. The grids are extruded in the streamwise (x_1) direction. The streamwise mesh spacing is set to $\Delta x_1^+ = 9.4$ according to findings of Gavrilakis [44] for DNS of square duct flow. These mesh properties are used for all subsequent cases if not explicitly highlighted differently, and are summarized in Table 7.1. Since the skin friction arising from the simulations deviates from Blasius law, the actual mean resolution differs from the initial estimation and will be given for the individual

Table 7.1: Standard domain size and spatial discretization where the viscous units are calculated using the fluid viscosity and the friction velocity estimated *a priori* with the Blasius correlation [106].

L_{x_1}/D_h	Δx_1^+	$\left(\Delta x_2^+\right)_{\text{wall}} \times \left(\Delta x_3^+\right)_{\text{wall}}$	$\left(\Delta x_2^+\right)_{\text{center}} \times \left(\Delta x_3^+\right)_{\text{center}}$
5	9.4	1.6×4.6	5×5

cases. In addition, the values of u_τ vary along the side walls of non-circular ducts. The normalization based on the local value at the most critical position provides the most conservative estimation of the grid resolution. The corresponding values are also provided for the different flow cases.
The total streamwise extensions are set to $L_{x_1} = 5D_h$ unless noted otherwise. This duct length corresponds to the set-up of Eggels *et al.* [32] for the DNS of circular pipe flow. However, the impact of the duct length on the result for non-circular duct flow is considered explicitly in Section 7.2.
As mentioned previously, the temporal resolution of the scales of turbulent flow is another important issue. To do this, the time step is chosen to ensure a Courant number of $Co_{\max} \leq 0.21$, which follows informations in the literature [44].

7.1.3 Initialization and evaluation of the turbulent flow field

Owing to the homogeneity of the flow in streamwise direction, periodic boundary conditions are used. The simulations are carried out under a constant flow rate condition, fixing the hydraulic Reynolds number in the simulations.
For the evolution of turbulent flow in the duct, appropriate initial conditions have to be defined. Various possibilities are known for how the flow can be disturbed in order to lead to the breakdown to turbulence [60]. However, as long as the transition to turbulence is triggered, the final turbulent flow field is memoryless with respect to the initial condition [34]. Within the present investigations, turbulent flow is initiated by a laminar parabolic profile with superposition of random disturbances $u_i' = 0.1\, U_b$ unless indicated otherwise. This procedure was also used by Faisst and Eckhardt for numerical investigation of the lifetimes of turbulent spots in pipe flow [34].
For the evaluation of the statistical properties of the flow, time averaging is performed for at least 40 turnover times, D_h/u_τ. Symmetries in the cross section and spatial averaging in the homogeneous streamwise direction are also used in order to speed up the convergence of statistics and to limit the computational costs.

The present approach certainly is a compromise between performing highly accurate turbulent simulations and using numerical tools for the development of a practically realizable flow control technique leading to energy savings in internal flows. It can be understood as an attempt to use a tool, which until now has been mostly restricted to scientific research for an investigation motivated by needs in engineering practice. This approach is only possible due to the considerable growth of computational resources within recent decades. The careful evaluation of the potentials and restrictions of the numerical procedure for the simulation of non-circular duct flows is felt to be of great importance for the reasonable interpretation of the results and will be discussed later. For this purpose, the turbulent flow in duct geometries that are known to show specific features of non-circular duct flows is investigated. The results are compared with findings in the literature in order to validate the present procedure for the aimed investigation of the distribution of the Reynolds stress field. It is noted, that studying the individual terms of the transport equation of the stresses (6.8) is not focused in the following and thus the ability of the present procedure for their computation is not evaluated.

7.2 DNS of turbulent flow in a square duct - comparison to literature data

7.2.1 Computation model

The turbulent flow through a straight square duct has been studied in direct numerical simulations, providing detailed reference data for comparison. The data obtained by Gavrilakis [44] and by Pinelli *et al.* [97] for $Re_h = 4410$ are considered in this respect. The former author uses a second-order finite difference approach for spatial discretization while in the latter investigation a spectral method is applied. Within these studies, the numerical results are validated by experimental data from different authors. Square duct flow includes the typical property of non-circular duct flows, *i.e.* the non-homogeneous distribution of the Reynolds stresses along the wall leading to the formation of secondary flow of Prandtl's second kind. Thus, this flow situation is considered to be a reasonable validation case for the present purpose.

Within the present study, the validity of the numerical method together with the chosen spatial and temporal discretization are considered. Further, the influence of the domain length on the computational result is analyzed, which has a significant impact on the numerical effort, *e.g.* the number of grid points that are required. Since the present investigation addresses the comparison of the flow through a number of differently shaped ducts, it is aimed at limiting the numerical cost

in this respect. From previous studies of circular pipe flow, it was found that the convergence of different turbulence statistics depends on the length of the computational domain [19]. The requirements for the pipe length are more critical for higher order statistics. According to Chin *et al.* [19], a duct length of $4\pi\delta$, where δ corresponds to the pipe radius, is sufficient for the convergence of the turbulent intensity components that will be studied in this work. However, owing to the special properties of non-circular duct flow, it is uncertain whether the same requirements hold here. For this reason, the results arising from square ducts with a streamwise extension of $5\,D_h$ (which approximately corresponds to the requirement for circular pipe flow) and $10\,D_h$ are compared.
As already mentioned, use of polyhedral cells for the discretization of complex duct shapes is particularly suitable. This choice is of course not logical for the discretization of a square duct. However, this duct shape offers the possibility of systematically investigating the influence of a polyhedral mesh on the computational result. The lower left quadrant of the meshes used within this study is shown in Figure 7.1.
A principal sketch of the computational domain together with the applied coordinate system is shown in Figure 7.2. The symmetries within the duct shape are used for the evaluation of statistical data. For the declaration of the characteristic positions, the lower left quadrant of the domain is used. The domain sizes, mesh types and grid sizes of the cases considered within this validation study together with the information for the reference cases are summarized in Table 7.2. Some differences in the mesh spacings are emphasized in particular. For the present meshes consisting of hexahedral cells, the streamwise grid size is increased compared with the standard set-up given in Table 7.1. In general, the mesh spacing used for the present simulations should be geared to that used by Gavrilakis [44] since the numerical methodologies are also related. The deviation provides insights into the requirements for the streamwise resolution and has to be kept in mind for the interpretation of simulation results.
Note that the wall units in Table 7.2 are calculated based on the fluid viscosity and the friction velocity u_τ estimated *a priori* with the Blasius correlation. However, since the values of u_τ vary along the side walls of non-circular ducts, the normalization based on the local wall-shear velocity gives the most realistic impression of the grid quality in the near-wall region. The grid resolution at the most critical points, namely at $x_3/H = 0.5$, is calculated based on this definition and given in Table 7.3.

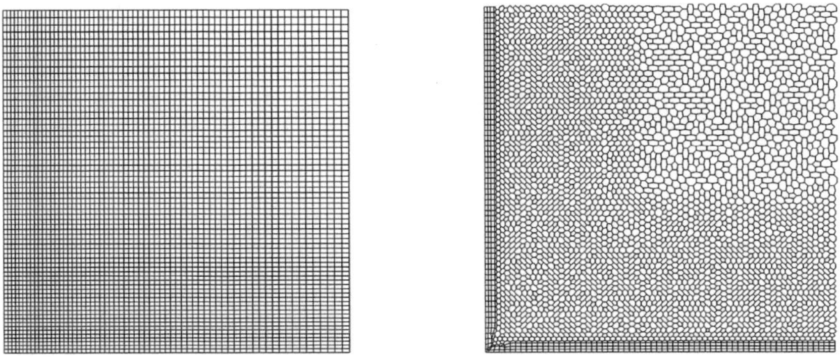

Figure 7.1: The lower left quadrant of the computational grids consisting of hexahedral (left) and polyhedral (right) cells used for the DNS of the turbulent square duct flow.

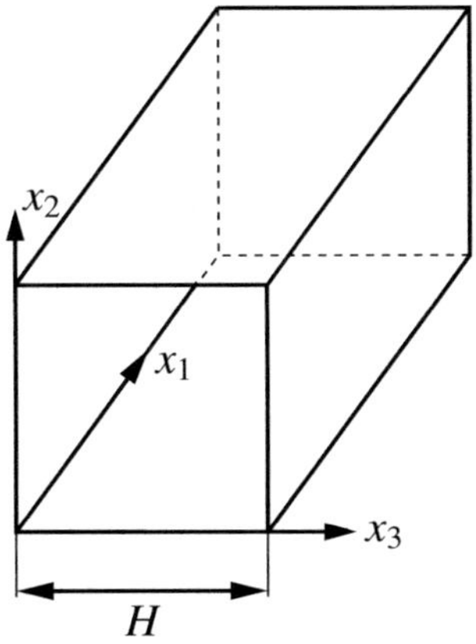

Figure 7.2: The computational domain for the simulation of turbulent flow in a square duct.

Table 7.2: DNS of turbulent flow in a square duct: domain size and spatial discretization in the current study and in the reference cases. The viscous units are calculated using the fluid viscosity and the friction velocity estimated *a priori* with the Blasius correlation [106].

case	cell type	Re_h	L_{x_1}/D_h	Δx_1^+	$\Delta x_{2,\,\mathrm{min}}^+$	$\Delta x_{2,\,\mathrm{max}}^+$
Current study						
hex, $L_{x_1} = 10D_h$	hexahedral	4405	10	15	1.6	3.3
hex, $L_{x_1} = 5D_h$	hexahedral	4405	5	15	1.6	3.3
poly, $L_{x_1} = 10D_h$	polyhedral	4405	5	9.4	1.2	5
DNS cases in literature						
Pinelli *et al.* [97]	hexahedral	4410	2π	14.7	0.07	5.2
Gavrilakis [44]	hexahedral	4410	20π	9.4	0.45	4.6

Table 7.3: Spatial discretization close to the wall at the most critical positions in the current study's cases with normalization based on the maximum friction velocity. The averaging time is expressed in terms of the mean wall shear velocity and the bulk velocity.

case	Δx_1^+	Δx_2^+	Δx_3^+	averaging time
hex, $L_{x_1} = 10D_h$	16.8	1.8	3.6	$73D_h/u_\tau$ or $1035D_h/U_b$
hex, $L_{x_1} = 5D_h$	17.1	1.8	3.6	$44D_h/u_\tau$ or $623D_h/U_b$
poly, $L_{x_1} = 10D_h$	10.2	1.3	2.6	$73D_h/u_\tau$ or $1082D_h/U_b$

7.2.2 Special properties of the velocity field

Theoretical considerations indicate that reduced energy dissipation in turbulent flows originates from a defined appearance of the Reynolds stresses as discussed in Chapter 6. Thus, the correct prediction of these quantities is essential in order to evaluate duct flows in the context of the analytical findings.
In Figure 7.3 the normal components of the Reynolds stress tensor resulting from the present simulations are compared with the results from Gavrilakis [44]. For this purpose, the development at two different spanwise positions is presented since the presence of side walls leads to a non-homogeneous distribution of the velocity field in the x_3-direction. Generally, a good agreement for all quantities can be observed for the different simulation cases. In particular, the data from the hexahedral meshes with different duct lengths generally collapse. This behavior suggests that the influence of the duct length is only marginal for the quantities analysed within this framework and a duct length of $L_{x_1} = 5D_h$ can be assumed to be sufficient.
However, if the development of the data is analysed in more detail, small systematic deviations can be observed: The magnitude of $u_{3,rms}$ is overestimated in the simulations with the hexahedral grids compared with the literature data. In contrast, the results produced form the polyhedral grid show a very good agreement with the data from Gavrilakis [44]. The development of $u_{2,rms}$ for the different meshes suggests that a similar trend than that observed for $u_{3,rms}$ appears, even if the corresponding literature data are not available for comparison. This behavior indicates a slight overestimation of turbulent activity with the hexahedral meshes and might be related to the increased grid spacing in the x_1-direction.
Finally, the results for $u_{1,rms}$ show very small deviations at the position $x_3/H = 0.35$, while they increase at the position closer to the side wall, $x_3/H = 0.15$. Obviously, the side walls have a strong impact leading to an increased complexity of the flow field. This behavior is also reflected by the larger deviations in the numerical results. The same trend is observed in a comparison of Gavrilakis' results with those of Pinelli *et al.* [97].
It can be shown analytically that the observed non-homogeneity of the Reynolds stress distribution along the duct walls leads to the well known peculiarity of turbulent flow through non-circular ducts, namely the formation of secondary flow of Prandtl's second kind [100, 104]. These counter-rotating vortices, which transport fluid from the duct center towards its corners, can be observed in all studies. A typical vortex pattern is shown in vector representation for the upper left quadrant of the simulation with hexahedral mesh and $L_{x_1} = 10D_h$ (see Figure 7.4). At the same time, the magnitude of the secondary motion, namely $\sqrt{\overline{U}_2^2 + \overline{U}_3^2}$ is

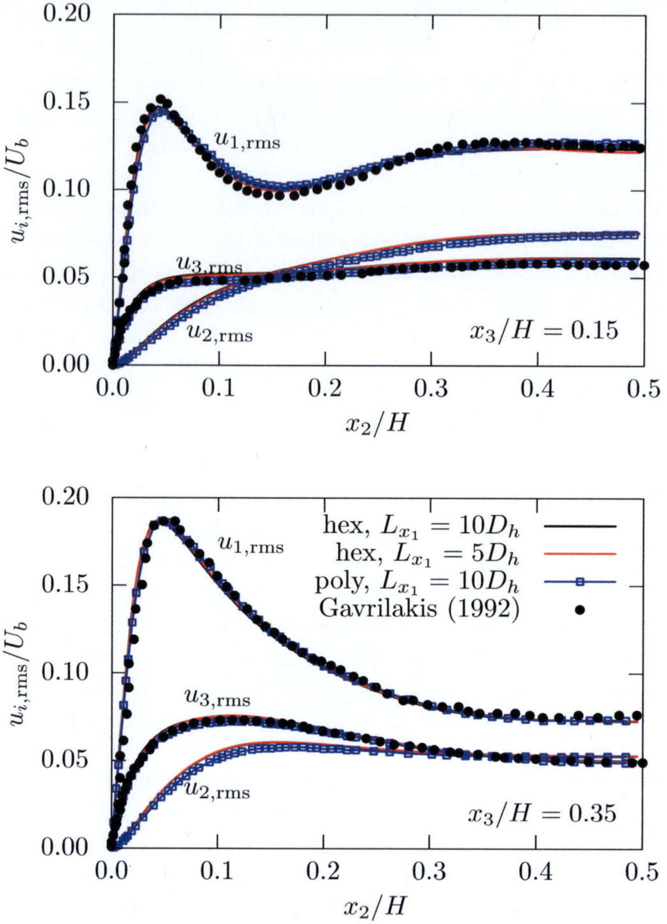

Figure 7.3: Development of the rms-values of the normal Reynolds stresses normalized with U_b at two different spanwise positions: $x_3/H = 0.15$ (upper figure), $x_3/H = 0.35$ (lower figure).

shown as a contour plot in the cross section area of the duct. In the reference work, the magnitude of the secondary flow is found to be maximal at the spanwise position $x_3/H = 0.23$ and reaches 1.9% of U_b [44]. Within the simulation with the polyhedral mesh, this value is reached fairly well. The results arising from the hexahedral mesh show a slight overestimation of the secondary motion, its magnitude reaches 2.1% of U_b. At the same time, the corresponding position moves closer to the wall to $x_3/H = 0.20$.
The secondary motion is studied in more detail in Figure 7.5, where the development of $\overline{U}_3$ at $x_3/H = 0.15$ is shown. This spanwise position is close to the place, were the magnitude of the cross flow reaches its maximum. It can be clearly seen that the velocity of the secondary flow is slightly overpredicted with the hexahedral meshes and again the duct length has only a minor effect on the computational result. The agreement of the data achieved with the polyhedral mesh with literature data is very good especially in the corner region up to $x_2/H = 0.15$, where the duct diagonal is reached. Obviously, $\overline{U}_3$ has a negative sign along the diagonal meaning that fluid is transported towards the duct corner.
Even if the magnitude of the secondary motion only reaches about 2% of U_b it influences the development of the mean streamwise velocity profile $\overline{U}_1$. In Figure 7.6, $\overline{U}_1$ is also shown at $x_3/H = 0.15$. At this position, the secondary motion provokes the velocity maximum to appear on the duct diagonal. This effect is met in all simulations. It is noted that the previously observed trend for the different meshes does not occur. The same holds for the development of $\overline{U}_1$ along the duct centerline shown in Figure 7.6. Obviously, the overall prediction of the mean streamwise velocity is less sensitive to the streamwise grid spacing than was previously found for other flow quantities.

7.2.3 The friction behavior

The special properties of the velocity distribution in non-circular duct flow strongly affects the friction behavior along the duct walls. In Figure 7.7 the wall shear stress distribution is shown for all simulations and normalized with the corresponding value calculated from the Blasius correlation. The non-homogeneity of the profiles along the wall is a feature of non-circular duct flow. In the duct corner the wall shear stress vanishes, while its maximum is reached in the middle of the wall. The appearance of secondary motion leads to additional local extrema in the distribution. These interacting effects are illustrated in Figure 7.4, where the normalized wall shear stress distribution is shown together with the secondary motion.
These general trends are visible in all data sets shown in Figure 7.7. However, differences in the friction behavior at the duct walls can be observed: For the simulations with hexahedral meshes an overestimation of the literature data along

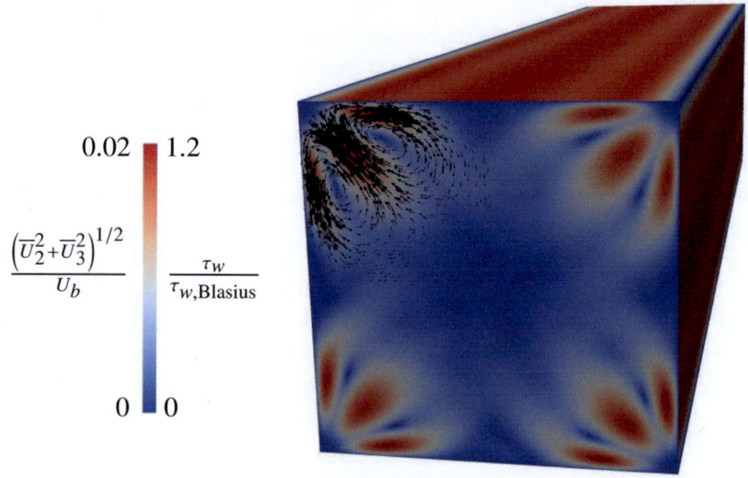

Figure 7.4: Illustration of the mechanism and the magnitude of the secondary motion together with the resulting distribution of the friction on the wall for the square duct.

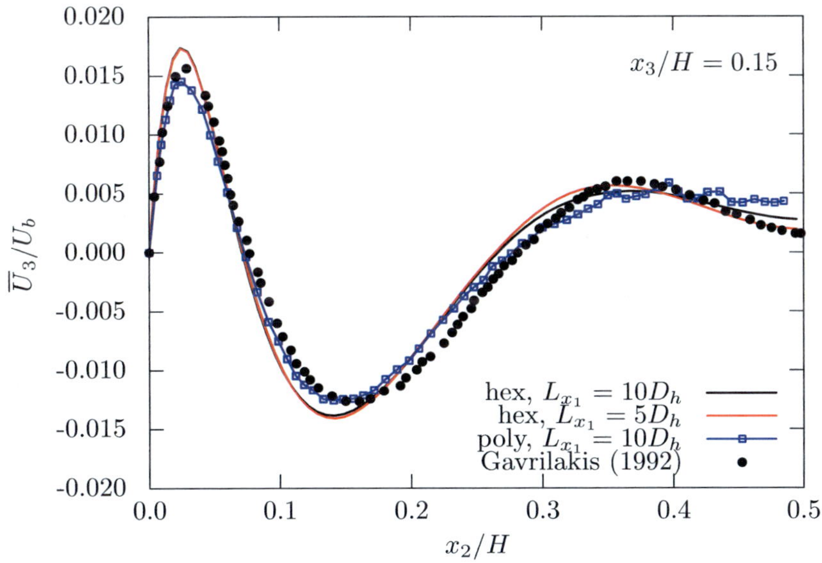

Figure 7.5: Development of the secondary flow component $\overline{U}_3$ at the spanwise position $x_3/H = 0.15$.

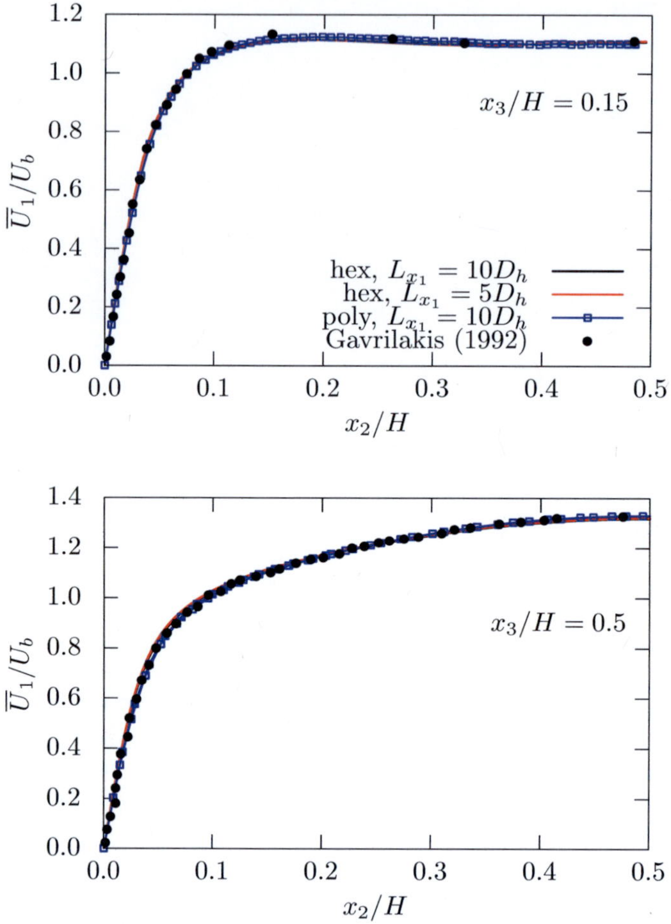

Figure 7.6: Development of the streamwise mean flow component $\overline{U}_1$ at the spanwise positions $x_3/H = 0.15$ (upper figure) and $x_3/H = 0.5$ (lower figure).

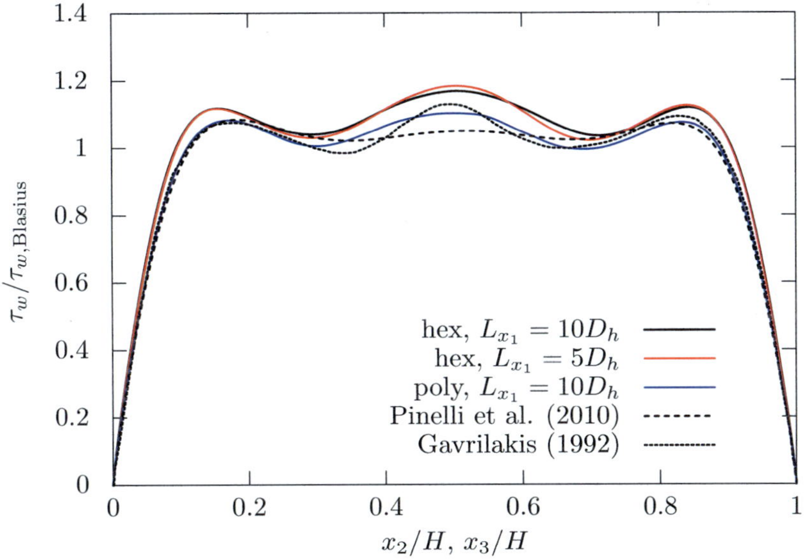

Figure 7.7: Development of the wall shear stress along the duct walls.

the entire duct wall is present. Again, basically no influence of the duct length is visible. In contrast, the data arising from the polyhedral mesh are in reasonable agreement with the findings from Gavrilakis [44]. The deviations around the center of the wall cannot really be explained, since the distribution is also different for the two reference cases in this region [44, 97].

Finally, from the average wall shear stress the dimensionless friction factor f can be evaluated according to Equation (4.3). It is shown for all simulations, together with the correlations of Blasius [106] and Jones [62], in Figure 7.8. The Jones correlation represents an improvement in the calculation of the friction in turbulent rectangular duct flows compared with, for example, the Blasius correlation. It is based on experimental data and is defined as

$$f^{-1/2} = 2\log_{10}\left(2.25\,Re_h\,f^{1/2}\right) - 0.8 \quad , \tag{7.1}$$

for square duct flow [62, 97].

Obviously, the friction losses resulting from the reference cases and the simulation with the polyhedral mesh are in very good agreement. All values are slightly lower than predicted by the Jones correlation, while the deviation is less than 2%.

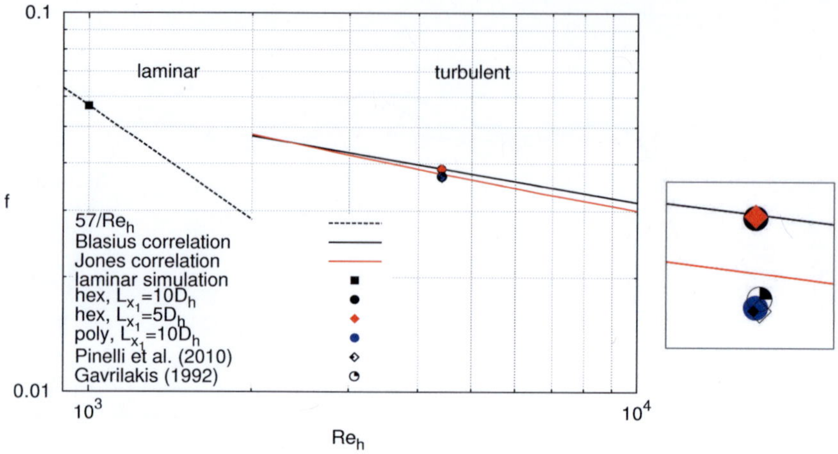

Figure 7.8: The friction factor resulting from square duct flow: comparison of analytical and empirical relationships with data from simulations with different grids and duct lengths.

In contrast, the friction factor for both simulations with hexahedral meshes agrees with the Blasius correlation and thus deviates from the reference data by about 5%.

The friction behavior observed in the different simulations confirms the observations made for the velocity distribution: for the cases containing hexahedral meshes, higher friction losses in comparison with the reference data appear due to an overestimation of the turbulent activity. These deviations might be related to the larger streamwise grid spacing in these cases and highlight the importance of sufficient resolution in this respect. At the same time, variations in the duct length do not influence the result significantly and the duct length $L_{x_1} = 5D_h$ is chosen for further studies. The flow field and the friction behavior arising from the polyhedral mesh shows very good agreement with reference data. These results illustrate that high quality DNS studies of non-circular duct flows can be performed using polyhedral meshes.

7.3 Simulation of the partial laminarization of turbulent flow

Besides the typical features of turbulent non-circular duct flow discussed in the previous section, an additional phenomenon is observed for a certain class of duct shapes. Experimental investigations by Eckert and Irvine [30] show the partial laminarization of turbulent flow in the proximity of the acute corner of a triangular duct with an 11.5° apex angle, leading to coexisting laminar and turbulent flow regions. In order to validate the numerical approach for prediction of this phenomenon, the DNS result of the flow in the duct geometry is compared with the few experimental data that are available. The outstanding properties of the flow field and the physical mechanisms involved in the laminarization process will be discussed in detail in Chapter 8.
The considered isosceles triangular duct with an apex angle of 11.5° is sketched in Figure 7.9. Additionally, the computational grid in the corner region, namely up to $x_3/H = 0.1$, is shown. The benefit arising from the usage of polyhedral cells is obvious: even if the corner region is very narrow, the appearance of critical cells that are skewed and have high aspect ratios is limited to the immediate vicinity of the tip. Additional details of the mesh and the computational domain are given in Table 7.4, while the viscous units are calculated using the fluid viscosity and the friction velocity, which is estimated *a priori* using the Blasius correlation [106]. The grid resolution expressed in terms of the ratio of the grid spacing to the Kolmogorov length, η_K, is another important issue for assessment the grid resolution in turbulence simulations. The value of the Kolmogorov length scale is estimated according to Lammers *et al.* [78],

$$\eta_K^+ = (0.25\, Re_\tau u_\tau / U_b)^{1/4} \quad , \tag{7.2}$$

and yields $\eta_K^+ \approx 1.5$ when the average friction velocity arising from the simulation result is applied. The mesh spacings expressed in terms of the Kolmogorov length scale are given in Table 7.5. In general, the quality of the mesh is considered to be satisfactory. However, the ability of the present numerical procedure to simulate the relevant scales of turbulent motion is studied in more detail. This critical assessment is performed for the simulation of the turbulent flow in the triangular duct. Similar results can be expected for the other flows which are studied.
In order to evaluate the adequacy of the spatial and temporal resolution of the DNS, the energy spectra of the frequency and the streamwise wavenumber are studied and are presented in Figure 7.10 and 7.11, respectively. These spectra are evaluated at the spanwise position $x_3/H = 0.7$ on the centerline of the duct where the flow is found to be fully turbulent. The expected slopes for the energy containing large

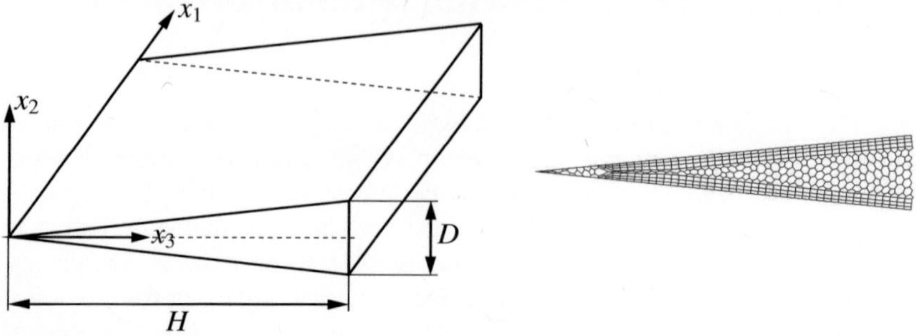

Figure 7.9: Sketch of the isosceles triangular duct with an apex angle of 11.5° where x_1 is the main flow direction (left) and the mesh in the corner region of the duct, namely up to $x_3/H \approx 0.1$.

Table 7.4: DNS of turbulent flow in a square duct: domain size and spatial discretization in the current study. The viscous units are calculated using the fluid viscosity and the friction velocity estimated *a priori* with the Blasius correlation [106].

cell type	Re_h	L_{x_1}/D_h	Δx_1^+	$\Delta x_{2,\mathrm{wall}}^+$	$\Delta x_{2,\mathrm{core}}^+$	$\Delta x_{3,\mathrm{wall}}^+$
polyhedral	4500	5	9.5	1.6	5	4.6

Table 7.5: Spatial discretization in the triangular duct with an apex angle of 11.5° based on Kolmogorov length scale.

Δx_1^+	$\Delta x_{2,\mathrm{wall}}^+$	$\Delta x_{2,\mathrm{core}}^+$	$\Delta x_{3,\mathrm{wall}}^+$
$5.7\eta_K^+$	$1.0\eta_K^+$	$3\eta_K^+$	$2.8\eta_K^+$

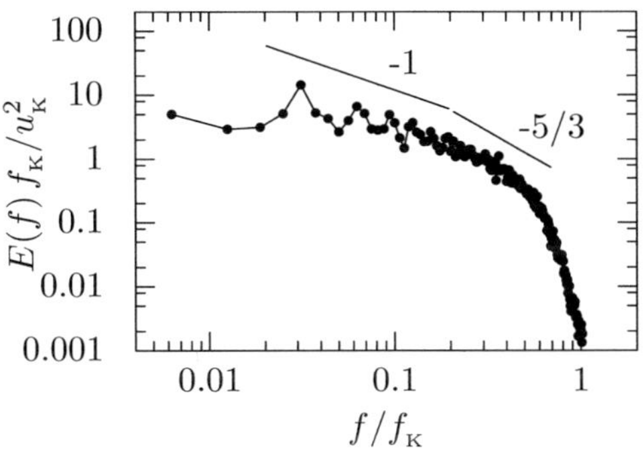

Figure 7.10: Frequency spectrum normalized with Kolmogorov scales arising at the position $(x_2, x_3) = (0.0, 0.7H)$.

scale motion (-1) and the inertial subrange $(-5/3)$ are also included in the plot. Owing to the normalization with Kolmogorov length and time scales, it can be observed that the flow is spatially and temporally discretized in a satisfactory manner: scales of $O(f_K)$ and of $O(\eta_K)$ are resolved. Additionally, the energy density associated with the high wavenumbers is several decades lower than the energy density corresponding to the low wavenumbers and there is no evidence of energy pile-up at high wavenumbers (see Figure 7.11). These criteria are discussed by Kim et al. [69] and indicate adequate grid resolution.
The available experimental data base compiled by Eckert [30] is restricted. However, the development of the mean streamwise velocity component along the symmetry line of the duct can serve for the validation of the present approach to predict flow laminarization. The data arising from DNS and from experiments are compared in Figure 7.12. In experiments, the part of the flow having laminar properties was found to be restricted to the corner region of the duct, namely to $x_3/H < 0.25$ [30]. The agreement of $\overline{U}_1$ in the laminar as well as in the turbulent flow region is considered to be acceptable since the experimental data might also be biased by inaccuracies in the measurement techniques.
Moving from $x_3/H = 0.25$ in a positive x_3-direction, the velocity profile was found to deviate from the laminar parabolic shape [30]. This behavior can also be observed in the calculated results. In Figure 7.13 the mean streamwise velocity component is plotted along the x_2-direction at the spanwise positions $x_3/H = 0.25$, $x_3/H = 0.3$ and $x_3/H = 0.5$. Parabolic shapes are included in order to explain the laminar properties of the flow. Obviously, at the position $x_3/H = 0.25$ the velocity distribution is parabolic. This behavior is lost when moving away from the corner.

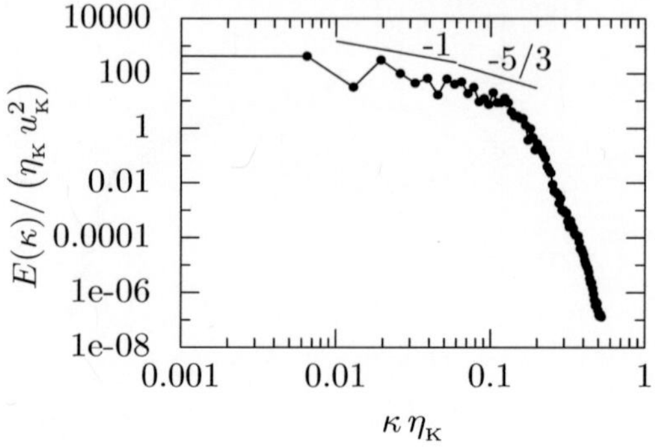

Figure 7.11: One-dimensional energy spectrum of the streamwise wavenumber normalized with Kolmogorov scales arising at the position $(x_2, x_3) = (0.0, 0.7H)$.

Thus, the existence and extension of the laminar region observed in experiments is also found in the numerical results. This leads to the conclusion that partial flow laminarization appearing in a certain class of duct shapes can be simulated with the present numerical approach.

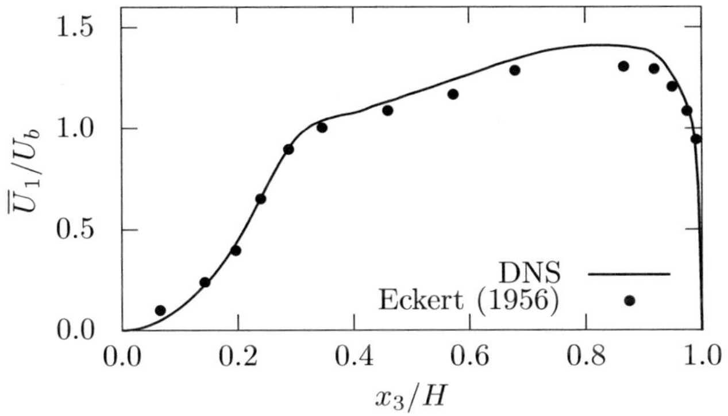

Figure 7.12: Development of the mean flow component $\overline{U}_1$ along the symmetry line of the triangular duct.

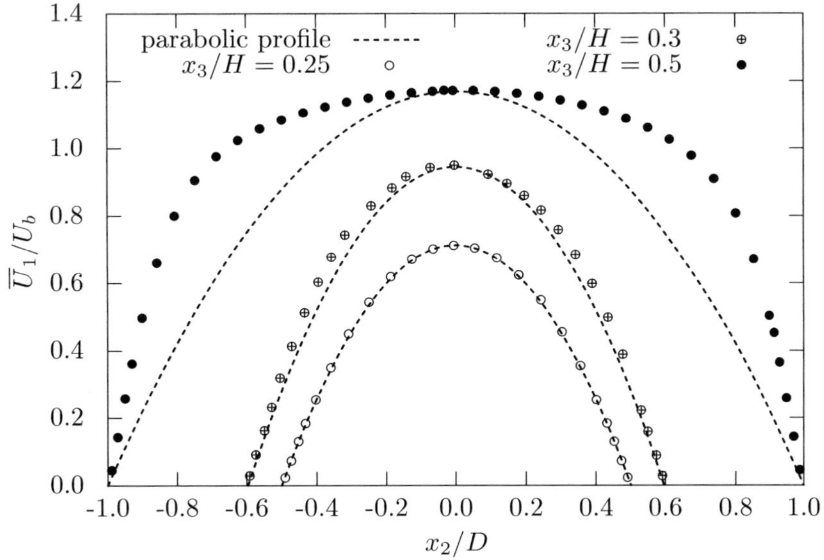

Figure 7.13: Development of the mean flow component $\overline{U}_1$ at the spanwise positions $x_3/H = 0.25$, $x_3/H = 0.3$ and $x_3/H = 0.5$ together with sketches of parabolic profiles.

8 Verification of the mechanism leading to low energy dissipation in duct flows

This chapter is dedicated to step **(3)** in the procedure for the development of energy-efficient duct shapes presented in Figure 6.7. In the theoretical considerations, particular statistical properties are suggested to characterize flow states with low energy consumption. Ducts leading to coexisting laminar and turbulent flow regions are considered for verification purposes. These flows show that geometrical properties can provoke laminarization of turbulent flow and enable the investigation of changes in the turbulent dissipation during the process. Most importantly, insights into the statistical flow properties accompanying reverse transition, transition and stable laminar flow are gained and a comparison with the theoretical findings can be made. In this respect, the investigation of coexisting flow regions arising in differently shaped ducts provides additional insights into the geometrical impact on the processes.

In order to investigate the coexisting laminar and turbulent flow fields, DNS are carried out with the method that was validated for this purpose in Chapter 7. The cross section shapes of the ducts that are considered in this chapter are shown in Figure 8.1. The shape in (a) represents a triangular duct with an apex angle of 12° and has been observed to result in the coexistence of laminar and turbulent flow in experiments performed by Eckert and Irvine [30]. In (b) a triangular duct with even smaller apex angle, namely 4° is shown, which has been investigated experimentally by Carlson and Irvine [18]. These measurements suggest that flow laminarization also takes place within this duct. Finally, a third duct shape was constructed, which is shown in (c) and will be referred to as a diamond shape. The ability of narrow corner regions to provoke the flow laminarization, which is observed for the triangular shaped ducts, is also expected for the converging angle in the corners of the diamond. All simulations are run at $Re_h = 4500$ as a significant laminar region and a turbulent region are expected under this condition. In Table 8.1 the mesh spacing is given at the most critical positions, namely where the wall shear velocity reaches its maximum.

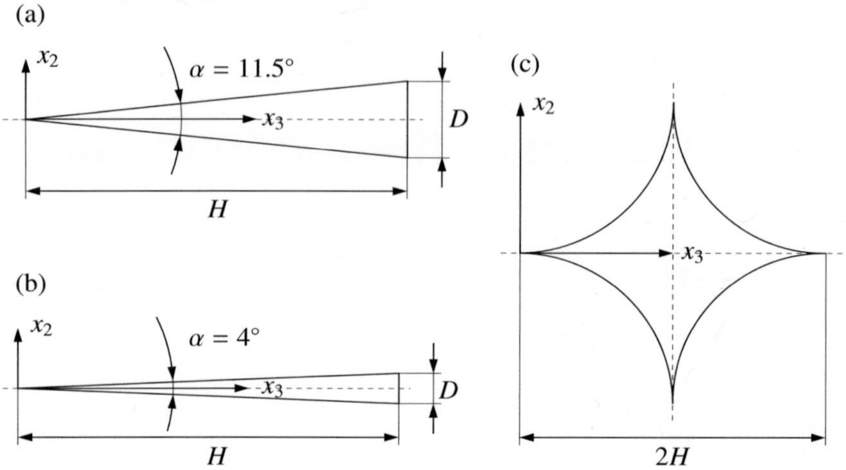

Figure 8.1: Cross section shapes provoking coexisting laminar and turbulent flow regions: (a) isosceles triangular duct with 11.5° apex angle, (b) isosceles triangular duct with 4° apex angle, (c) diamond shaped duct.

In addition, the averaging time employed for the statistical results that are discussed in the following are also listed.
In this chapter, the flow field arising from the duct shape (a) in Figure 8.1 is initially analysed and discussed within the context of experimental findings from Eckert and Irvine [30]. The anisotropy invariant representation of the stresses is used for the description of the variations in the flow field. This method allows a comparison with theoretical findings and can also be used to describe the transition process qualitatively. The findings for duct shape (a) are further compared with properties of the stresses appearing in duct shapes (b) and (c) (see Figure 8.1). Similarities and differences in the laminarization and the transition process are discussed.

Table 8.1: Spatial discretization close to the wall at the most critical position in the flow domains with coexisting laminar and turbulent flow fields. Normalization is based on the maximum friction velocity and averaging time is expressed in terms of the mean wall shear velocity and the bulk velocity.

case	position of $\tau_{w,max}$	Δx_1^+	Δx_2^+	Δx_3^+	averaging time
(a)	$x_3/H = 0.80$	10.9	1.8	5.3	$58 D_h/u_\tau$ or $920 D_h/U_b$
(b)	$x_3/H = 0.88$	11.3	1.8	5.4	$42 D_h/u_\tau$ or $690 D_h/U_b$
(c)	$x_3/H = 0.33$	10.2	1.6	3.0	$41 D_h/u_\tau$ or $740 D_h/U_b$

8.1 Flow in a triangular duct with 11.5° apex angle

Eckert and Irvine [30] investigate the flow through a triangular duct with an apex angle of 11.5° experimentally. Using flow visualization techniques, they observe the coexistence of a laminar and a turbulent flow within the cross section of the duct over a certain range of hydraulic Reynolds numbers where the flow is expected to be fully turbulent. The properties of these flow regions are further investigated in the following sections using detailed information of the flow extracted from the DNS. The appearance of a coexisting laminar and turbulent flow region further enables detailed studies of the transition process between the flow regimes. Based on theoretical findings, the anisotropy invariant representation of the stresses is expected to be a powerful tool for this purpose.

8.1.1 Properties of the coexisting laminar and turbulent flow field

Mean flow field

In order to provide a general impression of the mean flow field arising in the triangular duct with an 11.5° apex angle, contour plots of the mean velocity $\overline{U}_1$ and the magnitude of the secondary flow are shown in Figure 8.2 and Figure 8.3, respectively.
In Chapter 7, the flow field arising from (a) was compared with available experimental data for validation purposes. The laminar flow region was detected in the acute corner region of the duct up to approximately $x_3/H = 0.25$ in the present numerical investigation as well as in the experimental studies by Eckert and Irvine [30]. Moving further in the x_3-direction towards the core region of the duct, the mean flow field deviates from the parabolic laminar distribution and turns to the turbulent stage. In this part of the duct, secondary flows of Prandtl's second

kind [100] appear which are a characteristic phenomenon of turbulent flow through non-circular ducts: the counter-rotating vortices directed towards the corners on the right side are illustrated in Figure 8.3. It is observed that the magnitude of this secondary motion, namely up to 2.5% of the bulk velocity, is slightly larger than that in square duct flow, which Gavrilakis [44] found to be 1.9%. By contrast, the secondary motion directed towards the acute angle is very weak and vanishes when approaching the laminar part of the flow.
More details about the distribution of the mean streamwise velocity field can be gained in the comparison of streamwise profiles plotted along the x_2-direction at several spanwise positions, as shown Figure 8.4. The mean streamwise velocity is normalized with the corresponding local wall shear velocity, u_τ. In the following discussions the superscript "+" will refer to this type of normalization. For square duct flows such scaling was shown to be appropriate for the investigation of the near-wall region [44]. The characteristic profiles corresponding to the viscous sublayer, given by $U_1^+ = x_2^+$, and the logarithmic region, described by $U_1^+ = (1/\kappa)x_2^+ + B$ with $\kappa = 0.41$ and $B = 5.2$ [99], are also included in the plots.
Figure 8.4 (a) shows a velocity profile that is representative of the fully turbulent part of the duct. The result shows the expected linear relationship in the near-wall region. In the center of the duct U_1^+ approaches a logarithmic distribution, which is a common property of turbulent wall-bounded flows. The present data lie slightly above the relation given in [99] for plane channel flow. This observation is generally made for turbulent duct flows [32, 44]. When moving to positions closer to the corner shown in Figure 8.4 (b) and (c), typical deviation from the turbulent logarithmic distribution can be observed. The development towards a parabolic flow profile indicates the laminarization of the flow.

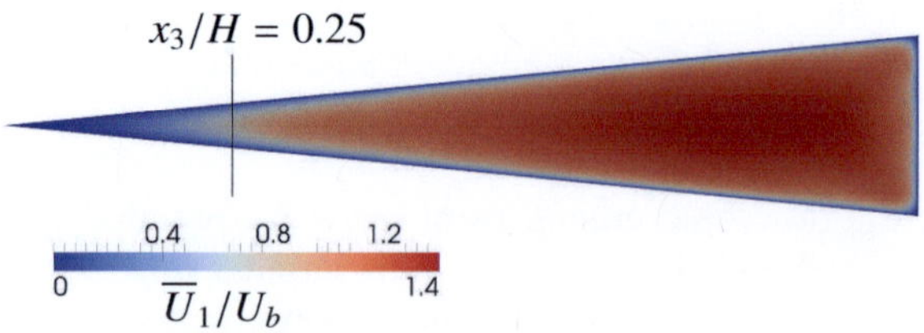

Figure 8.2: Contour plot of the mean streamwise velocity field $\overline{U}_1$ normalized with U_b.

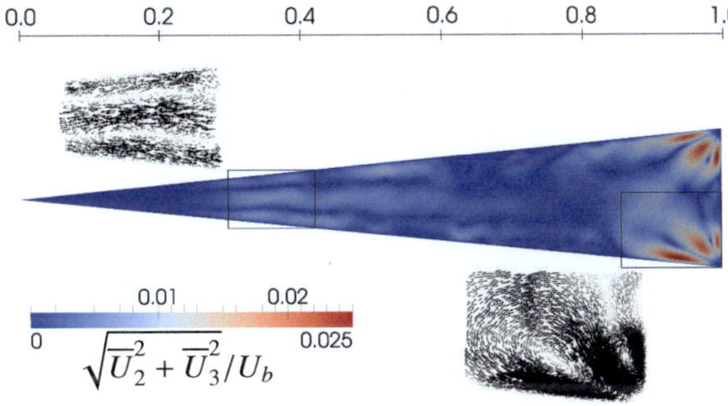

Figure 8.3: Contour plot of the magnitude of the secondary flow field $\sqrt{\overline{U}_2^2 + \overline{U}_3^2}$ normalized with U_b together with the secondary flow developing in the corner regions. Note that the vectors are scaled by the factor two close to the acute angle in order to distinguish the direction of the secondary flow in this part of the duct.

Scales and energy dissipation of the turbulent motion

Analysis of the energy spectra evaluated at three different spanwise positions highlight the changes in the scales of the turbulent motion when approaching the acute corner. The energy spectra of the streamwise wave numbers, $E(\kappa)$, is shown in Figure 8.5, where all quantities are normalized using Kolmogorov scales. These scales are estimated following Equation (7.2) and using the integral value of u_τ resulting from the simulation. In the turbulence-dominated region in the center of the duct corresponding to $x_3/H = 0.7$ and $x_3/H = 0.5$, the energy spectra collapse and indicate that the spectral transfer is being established towards features common for the turbulent energy cascade at relatively low Reynolds numbers. The energy containing, large scale motion illustrated by the slope "-1" is represented by a rather narrow band of wave numbers. The wave number range connected to the inertial subrange (slope "-5/3" in the plot) is also very restricted. This part of the spectrum represents important features of turbulence and becomes progressively extended for increasing Reynolds numbers [110]. Finally, at the high wave numbers, which correspond to the small scales of the turbulent motion, dissipation takes place.
The energy spectra at the position close to the acute corner do not show the development that is characteristic for turbulent flow. In particular, the wave number corresponding to the small scales is reduced, indicating a decrease in spectral separation in the flow.

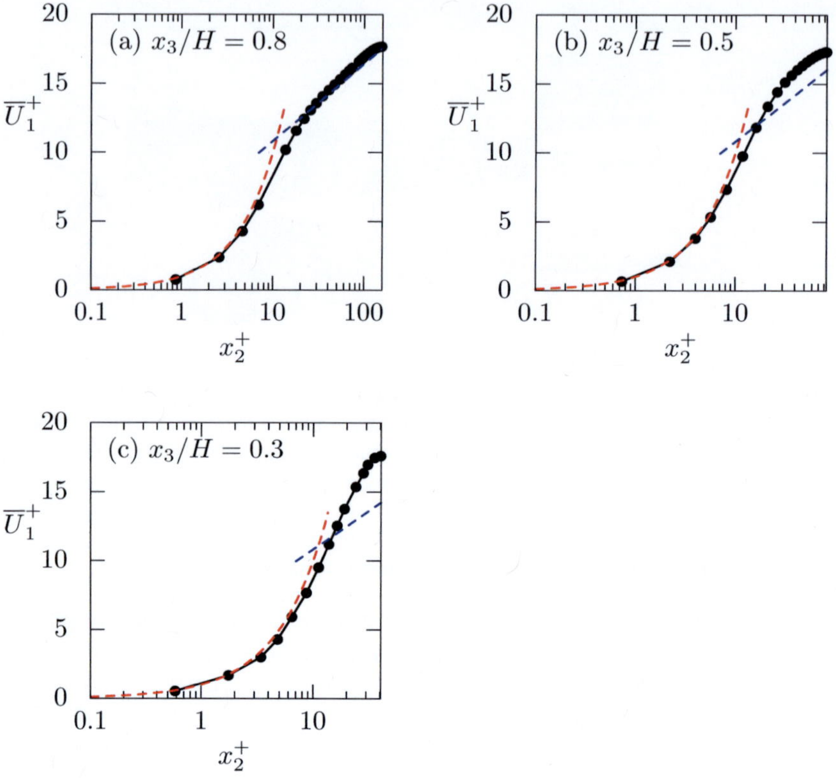

Figure 8.4: Mean streamwise velocity, $\overline{U}_1$, normalized with the local wall shear velocity, u_τ, plotted along the x_2-direction at different spanwise positions: $x_3/H = 0.8$ (a), $x_3/H = 0.5$ (b) and $x_3/H = 0.3$ (c), together with the characteristic profile for the viscous sublayer $\overline{U}_1^+ = x_2^+$ and the log-region $\overline{U}_1^+ = (1/\kappa)\, x_2^+ + B$ with $\kappa = 0.41$ and $B = 5.2$ [99].

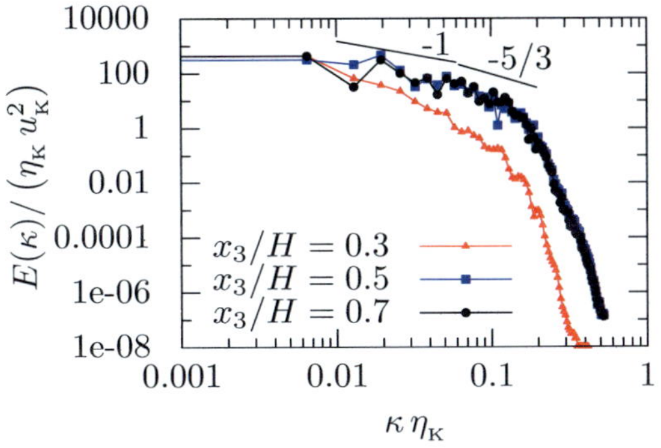

Figure 8.5: Wave number spectrum normalized with Kolmogorov scales at different spanwise positions on the x_3-axis of the triangular duct (a): $(x_2, x_3) = (0.0D, 0.3H)$, $(x_2, x_3) = (0.0D, 0.5H)$ and $(x_2, x_3) = (0.0D, 0.7H)$.

In Section 6.2.3 it was argued that decreasing spectral separation in the flow is associated with reduced turbulent dissipation. The validity of this statement for the flow behavior in the triangular duct is checked using the dissipation spectra arising at the same positions considered in Figure 8.5. The dissipation spectra can be evaluated from the wave number spectra, according to Tennekes and Lumley [110],

$$D(\kappa) = 2\nu\kappa^2 E(\kappa) \quad , \tag{8.1}$$

and are shown in Figure 8.6. Note that again Kolmogorov scales are used for normalization.

The integration of a dissipation spectrum over the entire wave number range yields a quantitative measure for the turbulent dissipation at the considered position in the duct cross section [110]:

$$\epsilon = 2\nu \int_0^\infty \kappa^2 E(\kappa)\,\mathrm{d}\kappa \quad . \tag{8.2}$$

Even if the dissipation spectra in Figure 8.6 show some scatter due to the restricted number of data samples, a clear tendency for ϵ can be extracted. While the turbulent dissipation is similar at the positions in the center of the duct, it is significantly reduced when moving towards the acute corner of the duct.

From this analysis, the connection between decreasing spectral separation and the energy dissipation of the flow becomes obvious. Moreover, it can be concluded that the duct corner provokes the aimed tendency towards reduced turbulent dissipation, resulting in $\epsilon \rightarrow 0$ and purely laminar flow in the close vicinity of the corner.

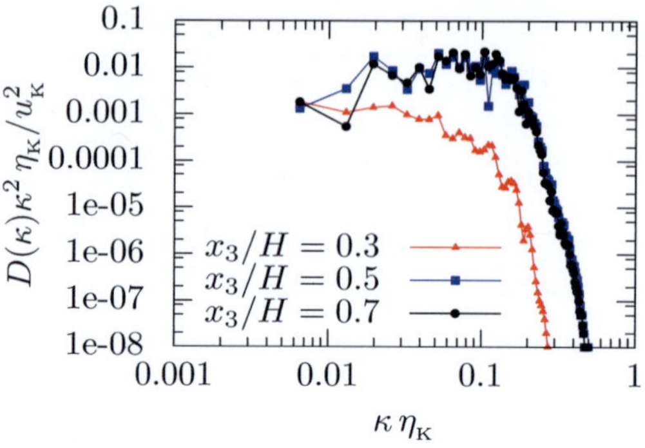

Figure 8.6: Dissipation spectrum [110] normalized with Kolmogorov scales at different spanwise positions on the x_3-axis of the triangular duct (a): $(x_2, x_3) = (0.0D, 0.3H)$, $(x_2, x_3) = (0.0D, 0.5H)$ and $(x_2, x_3) = (0.0D, 0.7H)$.

Statistical properties of the fluctuating motion

The statistical framework introduced in Section 6.1 is used in the following discussions to describe the suppression of turbulent flow properties towards the acute angle that accompanies the observed flow laminarization. In Figure 8.7 the dominant components of the Reynolds stress tensor are plotted along the x_2-direction at several spanwise positions. Owing to the very elongated duct shape in the x_3-direction some affinity to the profiles of plane channel flow are expected. These profiles for similar friction Reynolds numbers, $Re_\tau = u_\tau d/\nu$, are also included, where $2d$ is estimated by the width of the duct at a certain x_3-position. The exact values for Re_τ are: (a) $Re_\tau = 160$, (b) $Re_\tau = 85$ and (c) $Re_\tau = 40$, while channel data for $Re_\tau = 150$ (a) and $Re_\tau = 100$ (b) and (c) are used for comparison. The Reynolds stresses in Figure 8.7 (a) representing the core region of the duct in general are in good agreement with the corresponding channel flow data. The slight underestimation of the channel data in the near-wall region is also observed by Gavrilakis [44] for square duct flow and is related to the scaling with the local friction velocity, u_τ. At a position closer to the acute angle shown in Figure 8.7 (b) the magnitude of all stress components is considerably reduced compared with the turbulent channel data for the lowest Reynolds number available. This tendency increases when moving even further in the negative x_3-direction (Figure 8.7 (c)), where all components of the Reynolds stress tensor apart from $\overline{u_1 u_1}$ are almost suppressed, suggesting a highly anisotropic state of turbulence. Compared with channel flow data, a strong decrease of the maximum of $\overline{u_1 u_1}$ at $x_2^+ \approx 20$ in particular is observed. This peak in the streamwise stress component is characteristic of wall-bounded turbulent flows and accompanies the appearance of intense turbulent structures in the near-wall region [99]. The flattened maximum arising from the triangular shape of the duct indicates a strong damping of this near wall turbulent activity causing the laminarization of the flow.

Even if the magnitude of the stresses is significantly reduced when approaching the transition region, they still persist in the laminar part of the flow and are typically referred to as instabilities, disturbances or apparent stresses. The development of these stresses at two spanwise positions is shown in Figure 8.8. In contrast to the turbulent behavior discussed previously, these stresses reach their maximum in the center of the duct indicating the subcritical state of the flow [53].

In the close vicinity of the acute corner, only the streamwise stress component is present (Figure 8.8 (a)). When moving further in the spanwise direction, the spanwise component also increases in magnitude (Figure 8.8 (b)). This behavior is in agreement with findings from Lemoult *et al.* [79] for channel flow subjected to subcritical instabilities: for low Reynolds numbers they observe that the energy of the fluctuations is primarily related to the streamwise motion, while for higher Reynolds numbers spanwise fluctuations also gain in energy.

In addition to the observed increase in the streamwise and spanwise fluctuations when moving towards the turbulent part of the flow, these quantities appear to be correlated. This behavior indicates the initiation of production of turbulent kinetic energy.

The development of the stresses in the turbulent and laminar region of the duct can be further investigated using their invariant representation. This procedure is supposed to provide insights into the configuration of the stress tensor and enables the connection to the theoretical findings discussed in Section 6.2.

In Figure 8.9 the trajectories through the anisotropy invariant map are shown at different spanwise positions in the turbulent part of the flow. Obviously, the magnitude of anisotropy (II) at the wall and on the centerline increase when starting in the core region of the duct (Figure 8.9 (a)) and approaching the corner region (Figure 8.9 (c)). A closer look also suggests that the trajectories approach the right branch of the map representing the statistical axisymmetrical state of the stresses during this development. This observation is in agreement with the theoretically derived constraints for flow laminarization.

In Figure 8.10, trajectories that are representative of the laminar flow region within the duct cross section plane are drawn in the anisotropy invariant map. It is observed, that in the immediate vicinity of the duct, were the disturbances are found to be very small, only a narrow region in the invariant map is covered (see Figure 8.10 (a)). This region denotes a high anisotropy of the disturbances.

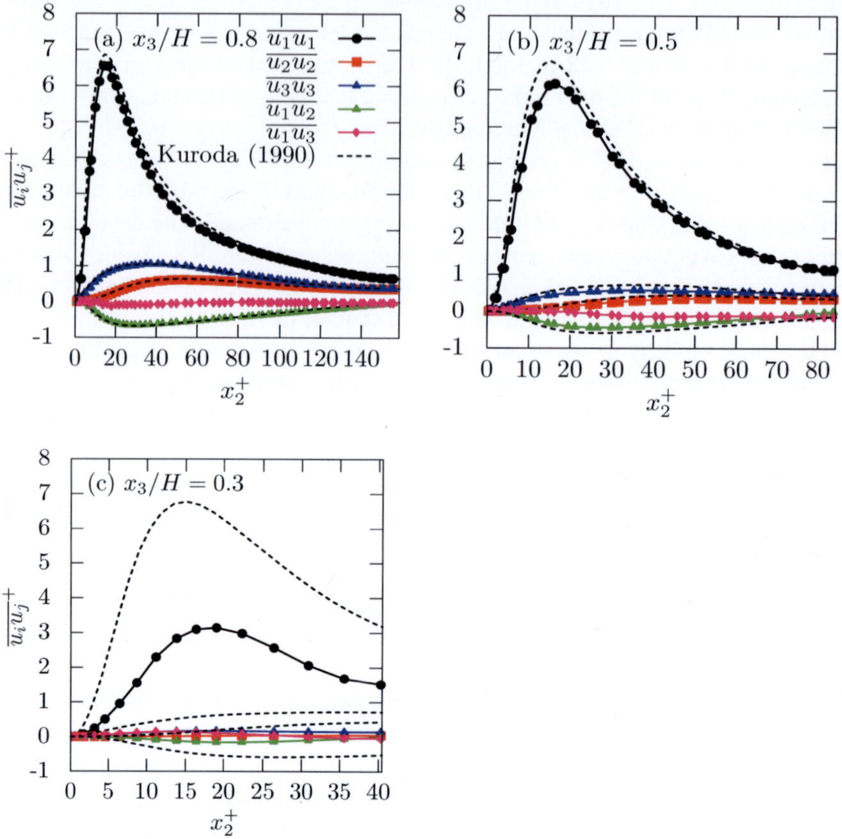

Figure 8.7: Reynolds stresses $\overline{u_iu_j}$ normalized with the local wall shear velocity u_τ plotted along the x_2-direction at different spanwise positions: $x_3/H = 0.8$ (a), $x_3/H = 0.5$ (b) and $x_3/H = 0.3$ (c), together with the profiles for plane channel flow at similar values of Re_τ. (a) $Re_\tau = 150$ [73, 74], (b) and (c) $Re_\tau = 100$ [73–75].

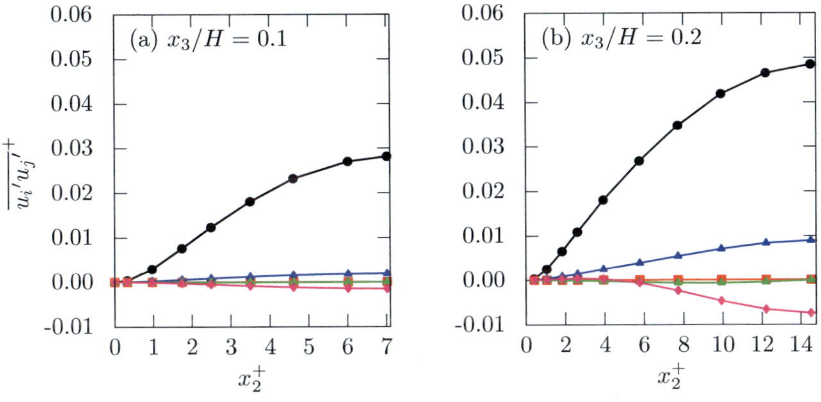

Figure 8.8: Apparent stresses $\overline{u_iu_j}$ normalized with the local wall shear velocity u_τ plotted along the x_2-direction at different spanwise positions: $x_3/H = 0.1$ (a) and $x_3/H = 0.2$ (b). Colors and symbols are used according to Figure 8.7.

When moving in the positive x_3-direction towards the transition to turbulence, the magnitude of anisotropy decreases (see Figure 8.10 (b)). The entire trajectory moves along the two-component limit of the map, indicating a significant deviation in the stresses from the axisymmetrical state. These tendencies conform to the theoretical findings since they clearly deviate from the conditions that are associated with stable laminar flow.
The analyzed invariant representation of the stresses at certain spanwise positions indicate an agreement with the theoretically derived constraints. In particular, the behavior of the stresses on the centerline of the duct is observed to show the characteristic trends. Thus, the trajectory along the centerline of the duct is assumed to enable a representative continuous description of the flow field.

8.1.2 Description of the transition process using the invariant representation of the stresses

So far, the properties of the flow in the laminar and the turbulent region appearing in the cross section plane of the duct have mainly been discussed separately. Since the present investigation focuses on discovering duct shapes that prevent turbulence and thus provoke the laminar flow state, a deeper understanding of the transition processes is essential.

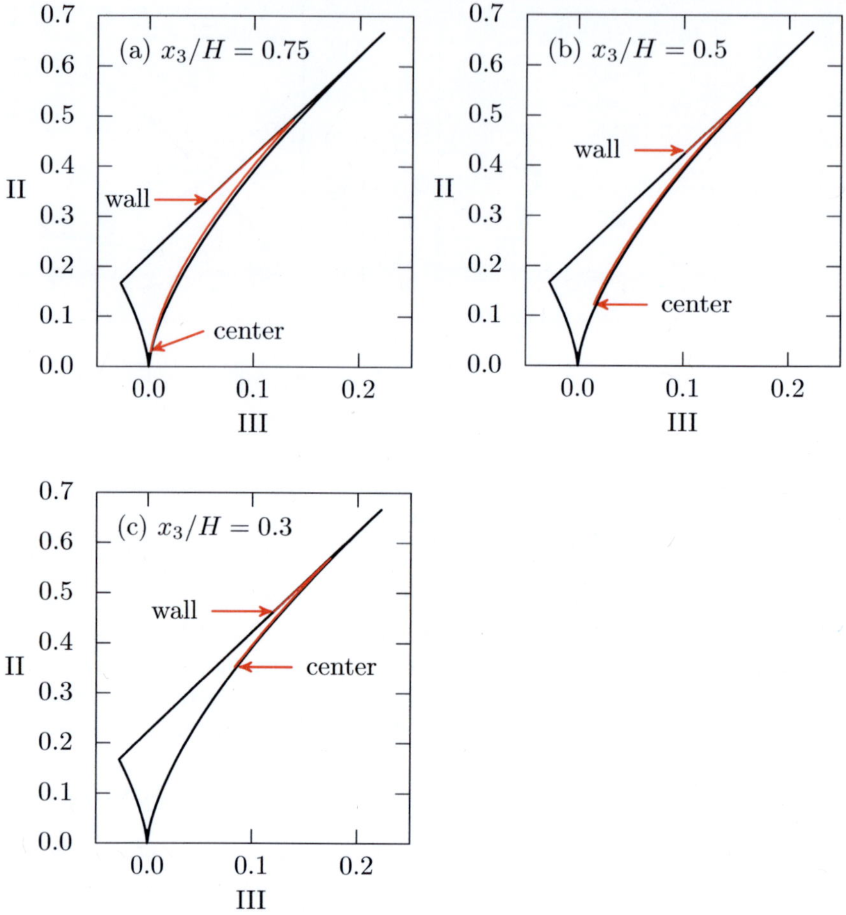

Figure 8.9: Trajectory through the anisotropy invariant map along the x_2-direction at different spanwise positions: $x_3/H = 0.75$ (a), $x_3/H = 0.5$ (b) and $x_3/H = 0.3$ (c).

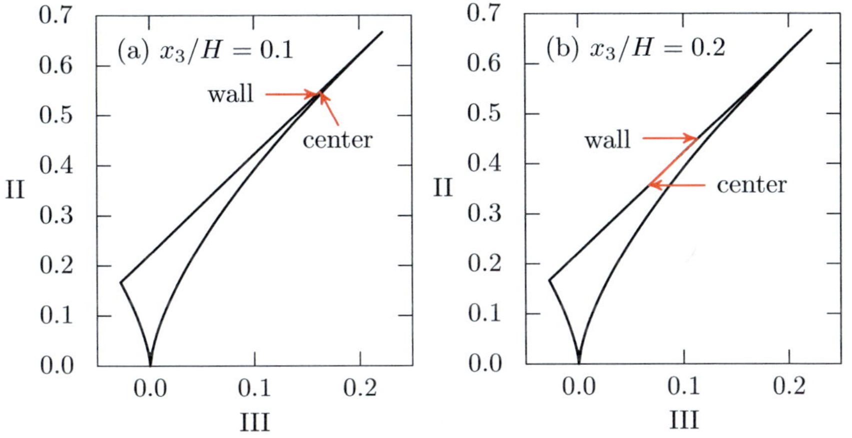

Figure 8.10: Trajectory through the anisotropy invariant map along the x_2-direction at different spanwise positions: $x_3/H = 0.1$ (a), $x_3/H = 0.2$ (b).

The present flow showing the coexistence of the two flow regimes enables mechanisms involved in the transition from laminar to turbulent flow, and *vice versa*, to be investigated.
In addition, the following analysis might also contribute to a fundamental area in fluid mechanics research. Even though transition has already been studied for many years, there are still plenty of unanswered questions concerning the phenomenon and the present statistical framework has rarely been used for its description.

Continuous description of the flow field

Based on the findings in the previous paragraph, the development of the stresses along the centerline of the duct cross section is expected to provide a continuous and characteristic description of the flow properties in the triangular duct [23]. The observation of decreased turbulent activity when moving towards the duct corner is confirmed by the development of the rms-values of the normal stress components along the x_3-axis in Figure 8.11. In the region close to the acute corner, the magnitude of all stress components is small, which is in agreement with the observed laminar flow behavior. Eckert and Irvine [30] use criteria in order to quantify the extension of the laminar and the turbulent flow region. These experimentally determined borders are also included in Figure 8.11: (i) represents the “instability line” [30] and (iii) is called the “transition line” [30]. When moving in the positive x_3-direction, Eckert and Irvine define the “instability line” at the position where they observe the first fluctuations by flow visualization. Passing this line the magnitude of the disturbances progressively increases. When reaching the “transition line” the mean flow profile starts to deviate from the laminar parabolic shape indicating the breakdown to turbulence. These experimentally proposed positions describing the transition process do partially coincide with outstanding positions in the development of the normal stresses extracted from DNS. The rms-values in Figure 8.11 are very small in the vicinity of the corner but stay almost constant when passing (i) up to $x_3/H \approx 0.20$. After reaching this position, first the magnitude of the streamwise stress component starts to increase significantly. When (iii) is reached, finally an increase in all rms-components can be observed indicating the breakdown to turbulence (compare also with Section 7.3). At the spanwise position $x_3/H = 0.36$, a local maximum of the streamwise rms-component is found, which is referred to as (iv). When moving further in the positive x_3-direction the fully turbulent core region of the duct is approached: at $x_3/H = 0.80$ the mean streamwise velocity component reaches its maximum. Passing this position the influences of the opposite wall become prominent and the stresses show the expected behaviour of turbulent wall-bounded flows. These effects are not the focus of subsequent discussions.

Until now, the transition from laminarity to turbulence was mainly observed to be associated with a significant increase in the magnitude of the Reynolds stresses (see Figure 8.7). The physical mechanisms that are involved in the transition process will now be investigated further. In Chapter 6 the methodology to describe the anisotropy tensor of the stresses in invariant space was presented, which illustrates the appearance of the stress tensor rather than the magnitude of its components. This tool offers the possibility to investigate the configuration of disturbances in laminar flow (“apparent stresses”) and of the Reynolds stress tensor in turbulent flow simultaneously. Based on the theoretical framework, it is expected that the invariant representation of the stresses represents a powerful tool for the investigation of the transition and reverse transition process since it is assumed to illustrate the relevant mechanisms. In Figure 8.12 the corresponding development from the corner to the core is shown in the invariant map and the resulting extrema are highlighted.

While the maximum of anisotropy, namely the one-component state of the stresses is reached within the corner ($x_3/H = 0.00$), the stresses in the core ($x_3/H \approx 0.80$) are almost isotropic. It is noted, that the latter spanwise position corresponds to the point where the maximal mean streamwise velocity is reached. However, the anisotropy in the stresses does not decrease monotonically from the wall to the

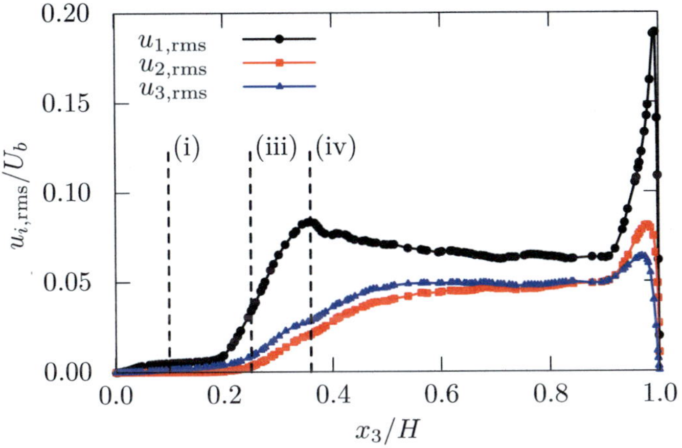

Figure 8.11: Development of the normalized rms-values of the velocity fluctuations $u_{i,rms}$ along the x_3-axis. (i) "instability line" at $x_3/H = 0.10$ [30], (iii) "transition line" at $x_3/H = 0.24$ [30], (iv) local maximum of $u_{1,\mathrm{rms}}$ at $x_3/H = 0.36$.

core but a rather complex trajectory through the anisotropy invariant map arises. In the corner region of the duct, the trajectory first develops along the two-component limit, which is associated with a strong deviation from the axisymmetrical state of the stresses. After the initial decrease in anisotropy, the development changes its direction at the position $x_3/H = 0.18$. The anisotropy in the stresses increases until $x_3/H = 0.24$ is reached, corresponding to the transition point found by Eckert and Irvine [30]. This behavior is followed by a monotonic decrease of II along the axisymmetrical border of the map.
The theoretical findings suggest that a highly anisotropic and axisymmetric state of the stresses causes reverse transition and stable laminar flow at the same time [23]. Qualitatively, an agreement of the trends in the trajectory through the anisotropy invariant map with these constraints is observed. However, a quantitative estimation, in particular of the deviation from the axisymmetrical state of the stresses, cannot really be made in this representation. In order to investigate the behavior of the flow in the triangular duct in more detail, corresponding quantities are proposed: the magnitude of anisotropy, II, and a measure for axisymmetry A, which is defined following the ideas of Antonia *et al.* [7] as

$$A = \frac{4/3|\mathrm{III}|}{(2/3\,\mathrm{II})^{3/2}} \quad . \tag{8.3}$$

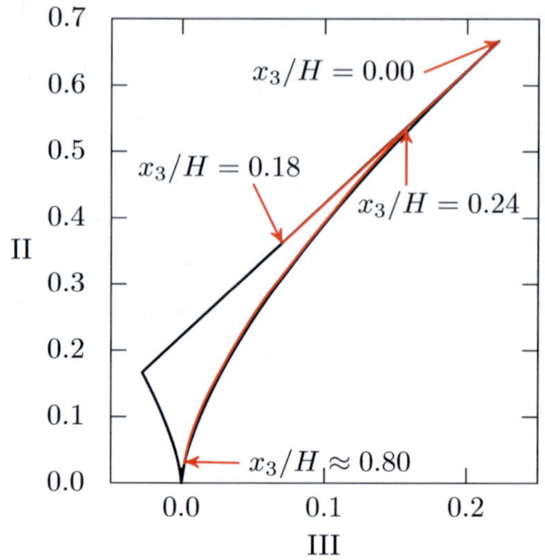

Figure 8.12: Trajectory through the anisotropy invariant map along the x_3-axis of the duct: extrema are marked and linked to the corresponding position in the flow domain (see Figure 8.1(a)).

$II_{max} = 2/3$ and $A = 1$ correspond to the maximum in anisotropy and axisymmetry in the stresses, respectively. The development of II/II_{max} and A along the centerline of duct is plotted in Figure 8.13. The outstanding spanwise positions for the magnitude of the stresses found in connection with Figure 8.11 are also included in the plot. It is clear that a correlation of the magnitude of the stresses and their configuration is present: the development of II/II_{max} and A also show an exceptional behaviour at the spanwise position (i), (iii) and (iv) (compare with Figure 8.11). The curves in Figure 8.13 highlight an additional spanwise position, $x_3/H = 0.18$, which is referred to as (ii).

Transition and reverse transition

Based on Figure 8.13, the trends in the configuration of the stresses within the transition from laminar to turbulent flow will be discussed first. For this purpose, the laminar region of the flow up to the transition line (iii) is investigated. In the direct proximity of the corner, the disturbances have an one-component shape, which agrees with the theoretical constraint for stable laminar flow. At the posi-

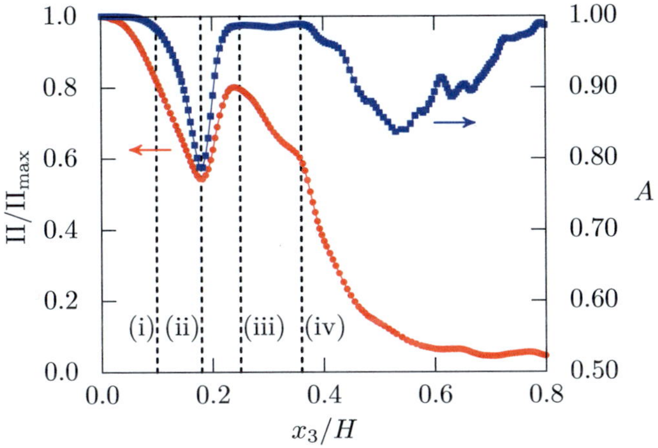

Figure 8.13: Development of the magnitude of anisotropy, $\mathrm{II}/\mathrm{II}_{\mathrm{max}}$ (red circles), and of the measure for axisymmetry, A (blue squares), along the x_3-axis. Additionally, characteristic positions within the transition process are indicated: (i) "instability line"at $x_3/H = 0.10$ [30], (ii) local minimum of A and $\mathrm{II}/\mathrm{II}_{\mathrm{max}}$, (iii) "transition line"at $x_3/H = 0.24$ [30], (iv) local maximum of $u_{1,\mathrm{rms}}$.

tion where Eckert and Irvine [30] observe the first disturbances in the flow, the stresses start to loose their initially axisymmetrical shape followed by a strong and continuous decrease of A and II. This deviation of the stress tensor from the theoretically derived stable state is provoked by fluctuations in the streamwise and spanwise direction. Even if these stresses are very small during this development their configuration describes the increasing instability of the flow. At $x_3/H = 0.18$ a local minimum in A and II is finally reached, which is marked by (ii). After this position, a significant increase of $\overline{u_1 u_1}$ can be observed in Figure 8.11 suggesting that the flow breaks down to turbulence. However, the final transition to turbulence in the mean flow field is not observed before (iii) (compare also with Figure 7.13). The discrepancy between these observations concerning the transition point will be further analysed in the next section.
Along with the transition from laminar to turbulent flow, the mechanisms involved in the process of reverse transition or flow laminarization can also be analysed based on Figure 8.13. In order to do so, this time the transition line (iii) is approached coming from the turbulent part of the flow and the discussion concentrates on the region $0.80 > x_3/H > 0.24$. When starting in the core region of the duct, where the flow is almost isotropic on the centerline, the movement in the negative x_3-direction first is accompanied by a deviation in the stresses from the axisymmetric state. This behavior is followed by a strong increase in A and II. It is observed, that the position where the local maximum of A is finally reached corresponds to the position where the local maximum of the streamwise rms-component is also reached (see (iv) in Figure 8.13). When moving further in the negative x_3-direction a significant increase in II is observed while A stays approximately constant. This process is associated with a drastic decrease of all rms-values (see Figure 8.11). At the position were II also reaches a local maximum (iii), the mean flow is finally found to turn into the laminar state. This behavior suggests that even if the axisymmetry in the flow is very high, its final laminarization does not take place before a certain anisotropy level in the flow is reached. However, it is noted that the magnitude of the stresses within the region, were they are highly axisymmetrical ($A \approx 1$), *i.e.* at the spanwise position $x_3/H = 0.3$, is strongly reduced compared with the turbulent channel flow (see Figure 8.7 (c)). The same holds for the turbulent dissipation (see Figure 8.6) indicating agreement with the analytically found effect of the axisymmetrical state of the stresses.

The border between laminarity and turbulence

A comparison of the present flow with fully turbulent circular pipe flow is used to detect the border between laminarity and turbulence. In Figure 8.14 the development of $\mathrm{II}/\mathrm{II}_{max}$ and A in the triangular duct is plotted starting from $x_3/H = 0.18$

or (ii) up to $x_3/H = 0.80$. In this plot, the corresponding data from the literature for fully turbulent pipe flow at $Re = 4950$ are also included [32]: the profiles are shown along the radius of the pipe starting at the wall ($x_3'/S = 0.00$) and reaching the center at $x_3'/S = 1.0$. The invariant properties close to the critical point (ii) in the triangular duct and within the viscous region of circular pipe flow show notable similarity. This behavior suggests that the turbulent flow in the triangular duct begins at the laminar corner layer in a similar way to that found for fully turbulent internal flows at the wall. Subsequently, (ii) is referred to as the "origin line".
The development of the mean streamwise velocity $\overline{U}_1$ along the centerline of the duct starting at (ii) can further support this finding. In Figure 8.15, $\overline{U}_1$ and x_3 are normalized using the friction velocity arising at (ii). A comparison with the expected linear behavior for turbulent flows close to the wall shows notable agreement in the vicinity of (ii). At a certain distance from the origin, the turbulence-dominated part of the flow is reached which is associated with a deviation from the initial viscous behavior. The region where viscous effects dominate seems to be extended compared with *e.g.*, pipe flow. However, it is noted that the flow field along the centerline of the duct is strongly affected by the diverging top and bottom wall resulting in a rather complex flow situation.
Based on the comparison of the stress configuration and mean flow in the triangular duct and the circular pipe it is found that the local minimum in $\mathrm{II}/\mathrm{II}_{max}$ and A indeed can be interpreted to be the border between the laminar and the turbulent flow region in the triangular duct. However, the viscous sublayer of the turbulent flow connects the two coexisting regions. The mean velocity in this layer has essentially laminar properties. Thus, when considering the mean flow field, the breakdown to turbulence cannot be distinguished in this part of the flow. This fact explains the differences in the critical points that are found when considering the mean or the fluctuating motion.
Despite the similarities of the flow in the triangular duct and in the circular pipe discussed here, differences appear when moving more to the central part of the ducts. These differences are assumed to originate from the fact that the triangular shape of the duct induces the laminarization of the flow. The process of reverse transition when approaching the critical point coming from the turbulent part of the flow is investigated in the following discussions. While the maxima in II and A are reached at similar radial positions in a circular pipe flow, A grows much faster than II in the triangular duct (see Figure 8.14). However, the connected process of laminarization is rather extended and the entire switch to laminar flow is accompanied by a significant increase in the maximum of anisotropy.

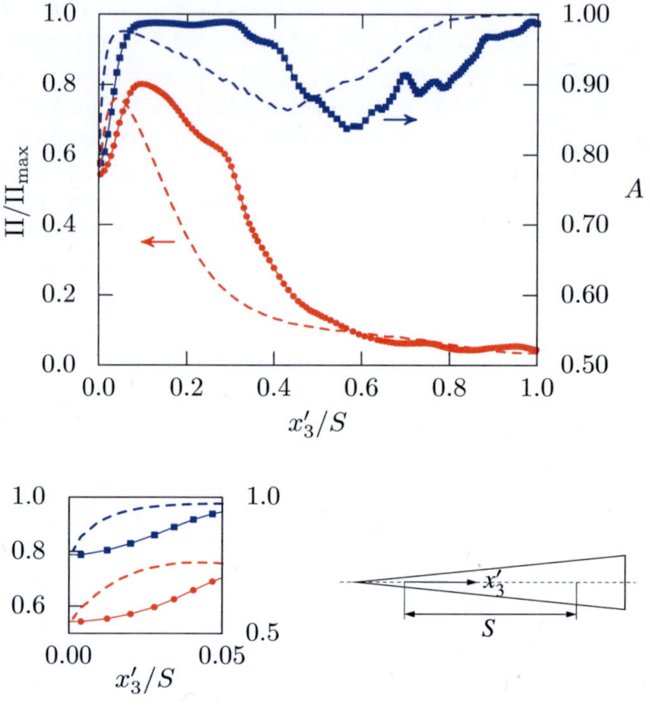

Figure 8.14: Development of the magnitude of anisotropy, II (red circles), and of the measure for axisymmetry, A (blue squares), along the x_3-axis of the triangular duct with 11.5° apex angle starting from the "origin line" (ii) and moving to the position were $\overline{U}_1$ reaches its maximum. II and A for circular pipe flow at $Re_h = 4950$ [32] are plotted with dashed lines.

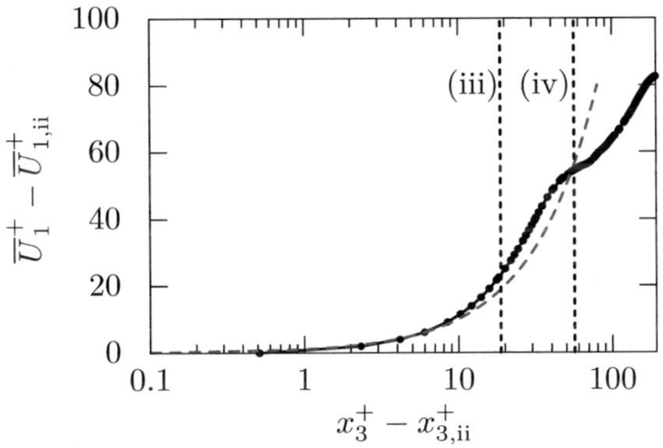

Figure 8.15: Development of the mean streamwise velocity, $\overline{U}_1$ along the x_3-axis of the duct. The profile starts at the "origin line" ($x_{3,\text{ii}} = 0.18H$) and ends at the point were the maximum value of $\overline{U}_1$ is reached ($x_3 \approx 0.8H$). For the calculation of the viscous units u_τ arising at the origin of the profile ($x_{3,\text{ii}}$) is used. Additionaly, the characteristic profile for the viscous sublayer, $\overline{U}_1^+ = x_2^+$, is indicated (red dotted line) together with characteristic positions within the transition process: (iii) "transition line" at $x_3/H = 0.24$ [30] and (iv) local maximum of $u_{1,\text{rms}}$.

8.2 Comparison of coexisting laminar and turbulent flow fields in related ducts

Along with the triangular duct with the 11.5° apex angle, additional geometries affecting the coexistence of laminar and turbulent flow were presented in Figure 8.1. A comparison of the observation made for the triangular duct with the 11.5° apex angle with the flow fields resulting from these geometries is expected to further support the understanding of conditions in the flow that characterize stable laminar duct flow [23]. Additionally, further insights into the constraints for flow laminarization are expected from this comparison. Finally, the impact of differences in the design of the corner regions on the transition scenario will be discussed.
In the previous section, for the triangular duct with the 11.5° apex angle, it was shown that the development of the magnitude of anisotropy, II, and the measure of axisymmetry, A, along the symmetry line of the duct describe the mechanisms involved in the transition from the laminar to the turbulent flow, and *vice versa*. This illustration is also used in the following analysis of related flow fields.

8.2.1 Characterization of stable laminar duct flow

The extension of the laminar region in triangular ducts was found to be a function of the apex angle, α. Bandopadhayay and Hinwood [10] worked out a model for its prediction, which is validated by experimental data from Eckert and Irvine [30] and Carlson and Irvine [18]. According to this model, the decrease in α results in an increase of the laminar region persisting next to the turbulent flow. The flow fields in the triangular ducts with 11.5° and 4° apex angles arising from DNS agree with this finding. While in the former case the laminar flow field extends to $x_3/H = 0.25$, the latter case shows a laminar flow behavior up to $x_3/H \approx 0.35$. Note that these are the spanwise positions after which the mean flow starts to deviate from the laminar parabolic shape.
The extension of the laminar flow region in the triangular duct with the 4° apex angle is confirmed by the magnitude of the rms-values along the x_3-axis in Figure 8.16. In the close proximity of the corner, all normal stresses are vanishing and only at some distance, $\overline{u_1u_1}$ start to grow followed by an increase in $\overline{u_3u_3}$. When the transition point (iii) is reached, finally the wall normal component, $\overline{u_2u_2}$, also increases denoting turbulent flow behavior. The data plotted in Figure 8.16 show some scatter in the region where the flow becomes unstable and breaks down to turbulence. The origin of this scatter cannot be finally clarified. Typically, the averaging time is expected to present an important parameter for the convergence

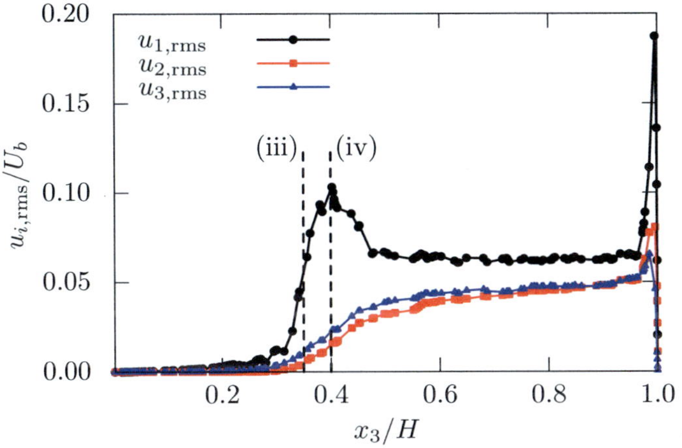

Figure 8.16: Development of the normalized rms-values of the velocity fluctuations $u_{i,\mathrm{rms}}$ along the x_3-axis of the triangular duct with the 4° apex angle (b): (iii) "transition line" at $x_3/H = 0.35$, (iv) local maximum of $u_{1,\mathrm{rms}}$ at $x_3/H = 0.40$.

of turbulence statistics. In Table 8.1, this period is given for the different duct geometries investigated. The scaling that is usually used for turbulent flows is applied. Obviously, the averaging time is comparable for all cases and was shown to result in a converged data set for geometry (a). However, the discussion that follows in Section 8.2.3 suggests differences in the transition scenarios arising in the duct geometries investigated, which might also affect the typical time scales of the flow and thus the time ranges that need to be considered for averaging (compare with Equation (6.2)). In spite of the present uncertainties, it is felt reasonable to discuss major trends in the data arising from geometry (b). It is also checked that these tendencies do not vary during the last $10D_h/u_\tau$ of the averaging period.

The increase in the spanwise extension of the laminar flow region, compared with the triangular duct with the larger apex angle, suggests that the flow in the latter geometry is more stable towards the amplification of disturbances. This behavior is further analysed using the invariant representation of the stresses. In Figure 8.17 the normalized magnitude of anisotropy, $\mathrm{II}/\mathrm{II}_{\mathrm{max}}$, and the measure of axisymmetry, A, are again plotted along the x_3-axis of the duct. In the stable laminar region of the duct, the flow is highly anisotropic and axisymmetric denoting the one-component state in the stresses. The agreement of this observation with theoretical constraints

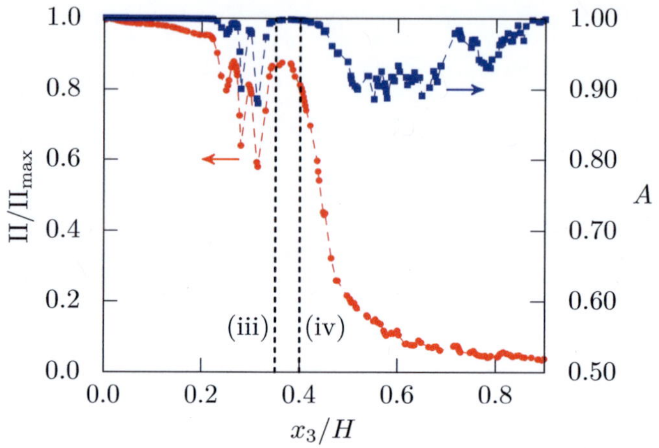

Figure 8.17: Development of the magnitude of anisotropy, II/II_{max} (red circles), and of the measure for axisymmetry, A (blue squares), along the x_3-axis of the triangular duct with the 4° apex angle (b). Additionally, characteristic positions within the transition process are indicated: (iii) "transition line"at $x_3/H \approx 0.35$, (iv) local maximum of $u_{1,rms}$ at $x_3/H = 0.40$.

for stable laminar flow indicates that the geometrical design of flow domains actually provoke the analytically proposed configuration of the stresses.

However, after a certain distance from the corner, the axisymmetry and anisotropy in the stresses decrease and the flow becomes unstable. This development does not follow the clear trajectory that was observed for the triangular duct with the 11.5° apex angle: II/II_{max} and A fluctuate over a certain distance, before the mean flow finally starts to deviate from its laminar properties at (iii). This fact most probably arises from convergence issues in the statistics, which were discussed previously.

8.2.2 Constraints for flow laminarization

In the following discussion, the intention is to provide a more general view of the mechanisms that are involved in the reverse transition process. For this purpose, observations made for geometry (a) in Section 8.1 are compared with the flow behavior in geometries (b) and (c) (see Figure 8.1 for references). This analysis is again based on the development of the normal stresses, the magnitude of anisotropy, II, and the measure for axisymmetry, A, along the x_3-axis of the ducts, which are shown in Figures 8.11, 8.13, 8.16, 8.17, 8.18 and 8.19. The geometrical symmetries

of the diamond shaped duct are exploited in order to improve the convergence of the statistical data set.
The normal stresses along the x_3-axis in the turbulent part of the flow show a similar behavior for all ducts: When moving from the center in the negative x_3-direction, first an increase of all components is observed while $u_{2,\mathrm{rms}} \approx u_{3,\mathrm{rms}}$ results in an almost axisymmetrical state. However, the magnitudes of the wall-normal and spanwise components decrease before the streamwise component reaches its local maximum (this spanwise position is referred to as (iv) in the considered plots). This behavior is connected to an increase of axisymmetry and anisotropy in the stress tensor. Actually, A reaches a local maximum at position (iv) in all ducts. The high axisymmetry in the stress tensor persists when moving from (iv) towards the transition line (iii). The continuous decrease of the normal stress components in this part of the flow indicates the reduction of turbulent activity. At the same time, a sustained increase in the anisotropy of the stresses is observed whose local maximum is reached at the transition line (iii).
Finally, it can be concluded that the reverse transition process indeed is characterized by high axisymmetry in the stress tensor. However, the final laminarization of the flow field does not take place before the stresses also are highly anisotropic. This behavior is similar for the ducts considered here, as is clearly emphasized for the duct shapes (a) and (c) in the inset of Figure 8.19.

8.2.3 Impact of corner designs on the transition scenarios

Within the previous paragraphs, the connection of the theoretical framework with the observations made for different coexisting laminar and turbulent flow fields was established. For this purpose, particular similarities in the these flows were detected and their physical meaning was discussed. The intention is now to concentrate on differences in the transition processes within these ducts in order to connect their distinct features to the individual corner designs. The following discussion is based on a comparative analysis of the development of the magnitude of anisotropy, II, and the measure for axisymmetry, A, along the centerline of the ducts shown in Figures 8.13, 8.17 and 8.19.
The impact of the corner design on the laminar to turbulent transition process is discussed first. The configuration of the disturbances in the laminar part of the ducts shows that the stable laminar region is large in the triangular duct with the 4° apex angle (b) compared with the triangular duct with the 11.5° apex angle (a) and the diamond shaped duct (c) when the duct height H is applied for normalization. A similar conclusion can be drawn for the regions where the disturbances become unstable and move along the two-component border in invariant space. In particular for geometry (c) the area where II decreases is shortened and the critical origin

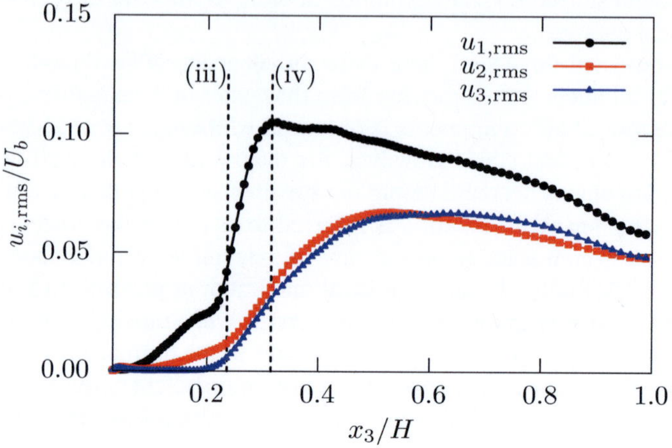

Figure 8.18: Development of the normalized rms-values of the velocity fluctuations $u_{i,rms}$ along the x_3-axis of the diamond shaped duct (c): (iii) "transition line" at $x_3/H = 0.24$, (iv) local maximum of $u_{1,\mathrm{rms}}$ at $x_3/H = 0.32$.

point (ii) is associated with a higher magnitude of anisotropy in the stresses than can be observed for geometries (a) and (b).

These differences in the flows can be quantified in terms of a transitional Reynolds number, $Re_{(\mathrm{iii})}$. It is estimated using the local height of the duct at the spanwise position (iii) as the characteristic length scale together with the bulk velocity appearing at this position. The results are summarized in Table 8.2. In the former discussion it was found that position (ii) in the flow domains describes fairly physically the border between laminarity and turbulence. The Reynolds numbers found at this position are also given in the table. However, since it is not possible to quantify $Re_{(\mathrm{ii})}$ for geometry (b) due to non-uniformities in the transition process, $Re_{(\mathrm{iii})}$ is used instead in the following analysis. The highest transition Reynolds number is observed for geometry (c). Interestingly this value matches the range that is found to be characteristic for plane channel flow, namely from 1100 to 1400 [17, 54]. The critical Reynolds number for channel flow is also given based on the bulk velocity and the entire channel height. It is noted that the geometry of the triangular duct with the 4° apex angle actually resembles a channel configuration with a large aspect ratio. Compared with this case, the transition Reynolds numbers resulting for geometries (a) and (c) are significantly reduced.

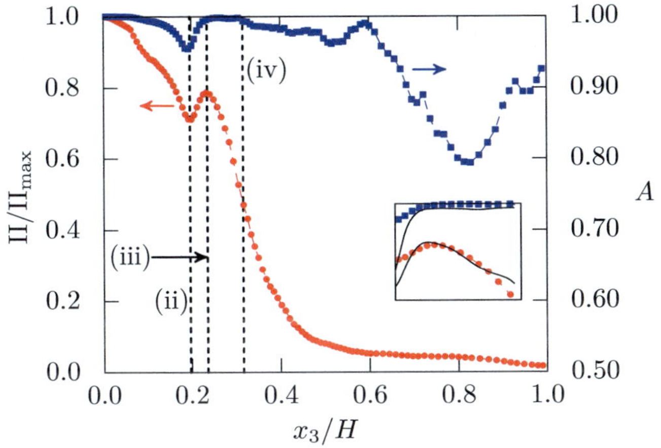

Figure 8.19: Development of the magnitude of anisotropy, II/II_{max} (red circles), and of the measure for axisymmetry, A (blue squares), along the x_3-axis of the diamond shaped duct (c). Additionally, characteristic positions within the transition process are indicated: (ii) "origin line" at $x_3/H \approx 0.20$, (iii) "transition line" at $x_3/H \approx 0.24$, (iv) local maximum of $u_{1,rms}$ at $x_3/H = 0.32$. The inset shows the development between the positions (ii) and (iv) in duct (c) together with the one in the corresponding region in duct (a) extracted from Figure 8.13 (black lines). Note that the scaling used in the inset is adapted to the data range considered.

This behavior might be connected to the individual appearance of mean secondary motion in the ducts: For the diamond shaped duct, which corresponds to the lowest $Re_{(iii)}$ the magnitude of the cross flow, $\sqrt{\overline{U}_2^2 + \overline{U}_3^2}$, reaches almost 3% of the bulk velocity, as can be seen in Figure 8.20. It is noted that this value is increased compared with the maximum value found for square duct flow at a similar Reynolds number, which corresponds to 1.9% [44]. In Figure 8.21 the magnitude of the secondary motion is also shown for a section of the triangular ducts, which is bounded by the transition line (iii) and extends to the center region were the scaling is adapted to the data range. It is clear that the magnitude of the secondary motion within these ducts, which only reaches about 0.8% of the bulk velocity, is significantly reduced compared with the diamond shaped duct. However, while a directed motion still is visible for the larger apex angle, it disappears for geometry (b). The weak cross flow that is visible for geometry (b) might be affected by

Table 8.2: Local Reynolds number arising at the origin point (ii) and at the transition point (iii) within the different duct geometries.

	case	$Re_{(ii)}$	$Re_{(iii)}$
(a)	triangular duct with 11.5° apex angle	210	600
(b)	triangular duct with 4° apex angle	-	1200
(c)	diamond shaped duct	30	120

the averaging issues discussed earlier. Thus, the strength of secondary motion in the turbulent region of the duct seems to be connected to the extension of the neighboring laminar region. Physically this fact might be explained by the convective properties of the cross flow: fluid with turbulent properties is directed towards the neighboring laminar region and limits its extension.

If the turbulent part of the flows, namely the region $x_3/H >$ (iii) is reconsidered in Figures 8.13, 8.17 and 8.19, it can be seen that the region of high axisymmetry in the stresses is largest for geometry (c) and decreases in (a) and further in (b). This tendency is similar to that observed for the strength of the secondary motion in the different ducts and an interconnection in the processes is suggested: the appearance of secondary motion is associated with highly axisymmetrical stresses on the centerline or corner bisector of a duct.

The observed impact of corner designs on the transition scenarios in coexisting flow situations shows that the strength of secondary motion can be linked to limits in the extension of the stable laminar region and to an accelerated laminar to turbulent transition process. However, at the same time the cross flow is accompanied with an enlarged region of high axisymmetry in the stresses in the turbulent part. Recalling findings from theoretical considerations in Section 6.2.3 and from simulations in Section 8.1.1, this appearance of the fluctuations can be linked to reduced turbulent dissipation. The interference of the secondary flow with the state of the stresses is not clear at this moment and its investigation is not the focus of subsequent discussions. Rather, it should be rather noted that the part of the flow where the axisymmetry is high is limited and the following proposal is made for the design of energy-efficient duct geometries: ducts consisting of corner regions that are shaped to extend the state of high axisymmetry in the stresses to large parts of the flow domain are suggested to promote the entire suppression of turbulent activity and thus enable large energy savings to be made.

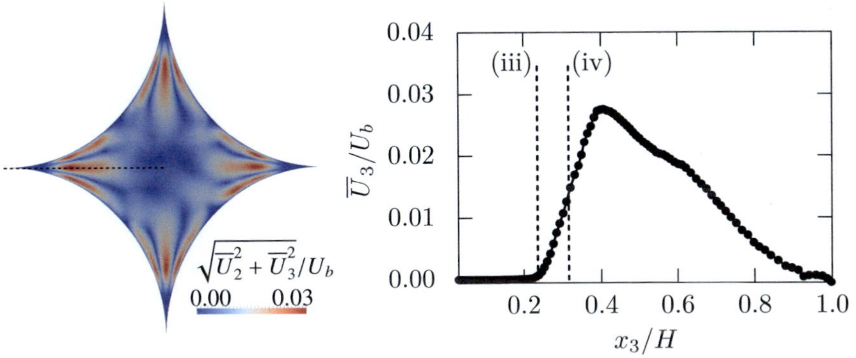

Figure 8.20: Contour plot of the magnitude of the secondary flow field $\sqrt{\overline{U}_2^2 + \overline{U}_3^2}$ normalized with U_b (left) together with the secondary flow $\overline{U}_3$ developing in a corner of the diamond shaped duct (right).

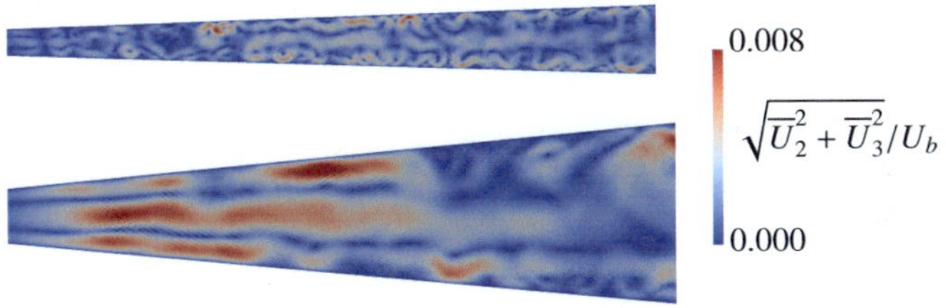

Figure 8.21: Contour plot of the magnitude of the secondary flow field $\sqrt{\overline{U}_2^2 + \overline{U}_3^2}$ normalized with U_b in a section of the triangular ducts with 4° (upper) and 11.5° apex angles (lower). The sections start at the transition point (iii) and end in the center region of each duct. Note that the scale of the color code is changed compared with Figure 8.14.

9 Impact of duct corners on turbulent flow

In the previous chapter, narrow corner regions in ducts were shown to locally provoke high axisymmetry and anisotropy in the stresses. This state is associated with stability of laminar flow and reduced dissipation of turbulent motion, which is followed by its laminarization. Generally, laminar flow is associated with significant lower friction losses than turbulence. Thus, ducts consisting of corner regions which can ensure laminar flow in situations where it typically is turbulent, are promising for the present optimization task of addressing reduced energy consumption of the flow (see Section 4.4).

In this chapter, the question to be discussed concerns which geometrical properties of corner regions in straight, non-circular ducts modify the flow in the theoretically proposed manner and how the effect can be expanded to large parts of the flow domain. For this purpose, the design of corners in the cross section plane and the impact on turbulent flow is investigated.

Turbulent flow in non-circular ducts has already been studied frequently in the literature in the past. These investigations mainly focused on the impact of several duct shapes on the frictional resistance and the mean flow field. First, a summary of these results will be given according to the historical sequence. In doing so, these findings are discussed in connection with the goals followed in this work.

Further, the available results are complemented with data from numerical simulations of novel duct shapes constructed of differently designed corner regions. These data enable the systematic investigation of how specific geometrical properties of the ducts influence the statistical flow field. Based on the findings in the previous chapters, these results are used to establish the potential of a certain property to promote the laminarization of turbulent flow and the delay of the laminar to turbulent transition.

9.1 Flow behavior in classical non-circular ducts

9.1.1 Review of findings in the literature

Friction resistance and energy dissipation

The impact of duct geometries on turbulent flow was studied intensively in the first half of the 20th century. Schiller [105] focused on the evaluation of the flow resistance of different types of non-circular ducts geometries, namely a square duct, a rectangular duct with an aspect ratio of 3.5, an equilateral triangular duct and a duct with a wavy boundary. These experiments investigated the validity of the concept of hydraulic diameter (see Equation (4.6)).
Schiller's results for the friction factor f of the equilateral triangular duct arising from these experiments are plotted in Figure 9.1 together with corresponding experimental data for the square duct [52] and isosceles triangular ducts with 11.5° and 4° apex angles [18, 31] (open symbols). The corresponding data from the present simulations are also included in the plot (solid symbols) and a good agreement for all duct geometries investigated can be observed.
In the laminar regime ($Re_h < 2300$) analytical solutions for the friction factor are available that obey the relationship given in Equation (4.7). For this flow regime the values emerging from the non-circular duct geometries are lower compared with the solution for the circular pipe. In contrast, the turbulent results for the square duct and the equilateral duct indicate that the scaling with the hydraulic diameter allows the adaption of the correlation describing the resistance of circular pipes to non-circular ducts with reasonable accuracy [106]. In fact, this law for the prediction of the flow resistance gives a good estimate for numerous duct shapes, *e.g.*, for elliptical ducts [91] and is widely used in engineering practice. In the present illustration, however, a deviating tendency arises for the considered data: while for the square duct the friction factor is only slightly reduced compared with the Blasius correlation, this effect becomes more pronounced for the triangular duct shapes. The tendency to lie below the empirical correlation of Blasius becomes stronger if the angle of the duct is reduced (Figure 9.1 (a)-(d)) [24]. The triangular shape with $\alpha = 11.5°$ (case (c)), for example, shows a constant reduction of f for about 20% over the entire turbulent regime investigated [30]. This fact may be explained using results for case (c) and (d) discussed in Chapter 8. There, the partial laminarization of the flow in these ducts was observed suggesting reduced friction losses compared with the fully turbulent flow state.
Moreover, it is noted that for the acute corners the transition from the laminar to the turbulent regime takes place at lower Reynolds numbers. This fact indicates

that the hydraulic diameter concept does not capture the underlying physics to generally unify the critical conditions leading to the breakdown to turbulence.

Secondary motion

Besides the evaluation of friction losses arising in non-circular ducts, characteristic structures of the mean flow field that are associated with a non-circular boundary are frequently discussed in the literature. In the 1920s, Nikuradse [92] performed pioneering work on the special properties of turbulent velocity fields appearing in

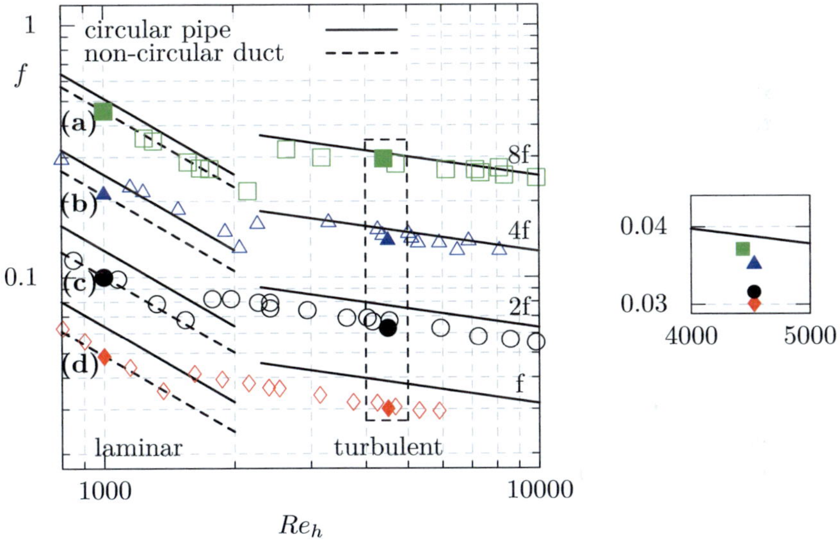

Figure 9.1: Friction factor f plotted against the hydraulic Reynolds number Re_h. Analytic solution for circular laminar pipe flow, $f = 64/Re_h$; Blasius correlation for turbulent flow, $f = 0.316/Re_h^{1/4}$ [106]; **(a)** Square duct: analytic solution for laminar flow, $f = 57/Re_h$; open symbols, measurements from Hartnett *et al.* [52]; solid symbol, DNS; **(b)** Equilateral triangular duct: analytic solution for laminar flow, $f = 53/Re_h$; open symbols, measurements from Schiller [92]; solid symbol, DNS; **(c)** Triangular duct ($\alpha = 11.5°$): laminar solution for a circular sector, $f = 50.3/Re_h$ [30]; open symbols, measurements from Eckert and Irvine [31]; solid symbol, DNS; **(d)** Triangular duct ($\alpha = 4°$): laminar analytic solution for a circular sector, $f = 48.85/Re_h$ [18]; open symbols, measurements from Carlson *et al.* [18], solid symbol, DNS. The right figure shows a comparative view of the DNS data at $Re_h \approx 4500$ (highlighted in the left picture).

straight non-circular ducts. He observed that the lines of constant mean streamwise velocity are displaced towards the corners at some distance from the centers of these ducts. Based on measurements of the streamwise flow field he and Prandtl were able to indirectly conclude the existence of secondary motion in form of streamwise vortices directed towards the corner of the duct for the first time [100]. Conclusively, the observed flow phenomenon is referred to as secondary motion of Prandtl's second kind. In Figure 7.4, the mechanism and the strength of the secondary motion arising in a square duct are shown in an example manner. From this illustration it becomes apparent that the corner vortices can be linked to the distribution of the wall shear stress along the duct sides and thus contribute to the friction behavior discussed in the previous section. In contrast to the phenomenon corresponding to the straight non-circular ducts considered here, secondary motion of Prandtl's first kind is pressure induced and appears in curved ducts of any cross section shape were centrifugal forces act at right angles to the main flow direction [27].
Owing to improvements in measurement techniques, the results of Nikuradse are followed by numerous experimental investigations involving direct measurements of the secondary flow field and the Reynolds stress tensor. These data fostered the physical understanding of the reasons for the appearance of secondary motion. In this context, Brundett and Baines [16] show that the production of streamwise vorticity is due to spatial variations of certain stress components, namely of $(\overline{u}_2^2 - \overline{u}_3^2)$ and of $\overline{u_2 u_3}$, in the wall-normal and spanwise direction. These experimental studies on secondary motion are extended by Gessner and Jones [46] who performed more detailed measurements of all Reynolds stress components. From these data they are able to determine on the dominant terms in the transport equation of the mean streamwise vorticity. Related studies for turbulent corner flow developing at approximately constant pressure and for turbulent flow in an equilateral triangular duct were conducted by Perkins [95] and Aly *et al.* [6], respectively.
In the early days of numerical simulation, scientists and engineers only depended on turbulence models for the calculation of turbulent flow fields. In the beginning, models were only able to handle isotropic turbulence and thus there were large restrictions in their applicability for the prediction of many flows of practical relevance. In later stages, these models were complemented to account for the anisotropic behavior of statistical flow fields and its spatial inhomogeneity. This improvement enabled the prediction of the flow field in non-circular ducts that frequently appears in engineering practice, *e.g.*, in heat exchangers, ventilation and air-conditioning systems, turbo machinery and open channels [27]. Corresponding work was done, for example, by Hanjalić and Launder [51] and Nakayama *et al.* [90] who tested their modeling approach for different duct geometries. A review

of experimental and modeling work on turbulent flow in straight non-circular ducts is given by Demuren and Rodi [27].
Owing to the increase of computational power, the DNS of the flow in a square duct at low turbulent Reynolds numbers, namely $Re_h = 4410$ or $Re_\tau = 300$, became possible. The results of Gavrilakis [44] finally give detailed information of the entire flow field, which are basically in agreement with previous experimental findings. Huser and Biringen [58] extended the data base from direct numerical simulation to a higher Reynolds number, $Re_\tau = 600$.
Further investigations in the present work follow the theoretical findings in Section 6.2 and concentrate on the impact of the duct design on the distribution of the stress field. The appearance of secondary motion is not explicitely discussed in this context. However, the results from DNS and experiments that were summarized earlier show that this property of the mean flow originates from the stress distribution, and thus can be assumed to be captured by the present thinking.
Besides the secondary motion in the mean flow, quasi-coherent structures can be identified in turbulent flow by conditional averaging. The interaction of the arising near-wall streaks and the appearance of larger scale streamwise vortices, namely secondary motion of Prandtl's second kind, is recently discussed by Pinelli *et al.* [97] using DNS data. Subsequently, these authors also investigate the flow in a square duct at a marginally turbulent state [111]. Within this work, the characteristic coherent structures appearing during flow laminarization are discussed. Starting from a fully turbulent state, it is found, that flow laminarizes at a Reynolds number of about $Re_h = 2154$. This Reynolds number is higher than the critical value found for channel flow but smaller than that of circular pipe flow. The difference from channel flow is discussed in the context of the side walls that limit the extension of the coherent flow structures in the spanwise direction. It is found that flow laminarization is connected to walls that are shorter than 154 wall units.
The results from Uhlmann *et al.* [111] support the present approach in two major aspects. Firstly, their findings suggest that geometrical modifications leading to spanwise limitations may interact with the transition behavior of the flow. Further, they find that the required scale of these limitations is rather large and significantly increased compared with that of, for example, riblets that act in turbulent flow (compare with Section 6.2.4). The underlying reasoning for the interaction of coherent structures with flow laminarization is not focused on here. However, an outlook for related interpretation of the data that are presented in the following sections is given in Section 9.2.3.

9.1.2 Properties of the turbulent stresses

Following along the path of this work, the statistical properties of the flow in the square duct and an equilateral triangular duct are investigated. Since these data are only partially available in literature, the results from simulations carried out during the course of the present work are used. In order to gain further insights into the impact of the different duct geometries on the turbulent stresses, their development along the corner bisectors with length R_b and along the wall bisectors with length r_b (see Figure 9.2) are discussed.

In Figure 9.3 the development of the Reynolds stresses along the wall bisector and the corner bisector of the square duct are shown. In order to highlight the special properties of the stresses in the non-circular duct, the corresponding data for circular pipe flow and plane channel flow are also shown [5, 32]. Note that the stresses in the cylindrical coordinate system of the circular pipe are adapted to the cartesian system using $u_z = u_1$, $u_r = u_2$ and $u_\varphi = u_3$. Firstly it is observed that $\overline{u_1 u_1}$ is increased at some distance from the wall of the circular pipe compared with the plane channel, while $\overline{u_2 u_2}$ and $\overline{u_3 u_3}$ are decreased, leading to an increase in anisotropy of the stresses in the former flow. This tendency is even more pronounced for the flow in the square duct along the wall bisector (upper plot in Figure 9.3). The observation made for the square duct also results from the scaling with the overall mean wall shear velocity arising from the simulation. It is shown by Gavrilakis [44] that a scaling with the local value of u_τ leads to a reasonable agreement of the stresses along the wall bisector and in channel flow at some distance beyond the viscous sublayer. However, since the present intention is to discuss the statistical properties of the flow in the non-circular ducts at different

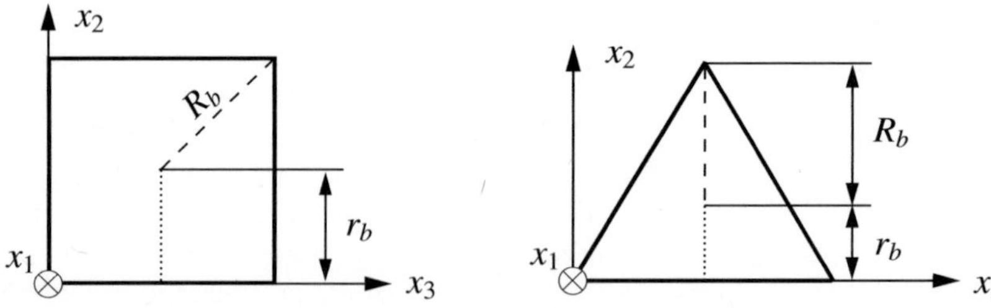

Figure 9.2: Cross section plane of the square duct (left) and equilateral triangular duct (right) with corner and wall bisectors. The wall bisectors with the length r_b are shown as dotted lines, the corner bisectors with the length R_b are shown as dashed lines.

positions in a general frame, scaling with the mean wall shear velocity is felt to be an appropriate choice. This scaling allows conclusions about the relative changes in the flow within a certain duct geometry to be made. Thus, it is possible to identify at which position the magnitude of turbulent stresses is increased or decreased.

Compared with the observations made previously, the flow along the corner bisector of the square duct (lower plot in Figure 9.3) shows a distinct behavior: due to the geometrical properties of the duct, the stress tensor appears to simplify since $\overline{u_2^2} = \overline{u_3^2}$ and $\overline{u_1 u_2} = \overline{u_1 u_3}$ suggesting an almost axisymmetrical state of the flow [16]. This state is associated with a significant decrease of all stress components up to a considerable distance from the wall compared with the wall bisector and the channel and pipe flow. Close to the corner of the duct, all stress components are damped leading to the disappearance of turbulent properties in the flow.

The impact of the corner angle on the flow behavior can be studied if the flow in the square duct is compared with that arising in an equilateral triangular duct. The corresponding distribution of the stresses along the wall and the corner bisectors of these ducts are shown in Figure 9.4. It is clear that the normal stresses along the wall bisector are increased (see upper plot in Figure 9.3) for the triangular duct. This behavior can be interpreted as enhanced turbulent activity in the area of the wall bisector. The shear stresses, however, are only slightly affected by the geometrical properties. For the square duct $\overline{u_1 u_3}$ and $\overline{u_2 u_3}$ vanish along the wall bisector due to the geometrical symmetry of the duct [16]. A similar behavior is observed for the triangular duct by Aly *et al.* [6] and in the present simulations.

In the lower plot in Figure 9.3, the stresses developing along the corner bisector are compared. The tendency towards damped turbulent activity in the corner is increased for the smaller angle of the triangular duct. In particular, the region in which the turbulent stresses entirely vanish is extended compared with the square duct. Additionally, the development of $\overline{u_1 u_1}$ is fairly untypical: in general, the streamwise stress component is decreased compared with the square duct and no characteristic peak value appears.

The configuration of the stresses in the square duct and the equilateral triangular duct is further investigated in the anisotropy invariant map. In Figure 9.5 the trajectories along the wall and the corner bisectors are presented. In general, the behavior in the two different ducts is somewhat similar. Along the corner bisectors (Figure 9.5 (b) and (d)) the one-component state is reached at the wall and an overall strong tendency towards axisymmetry is observed. Differences in the geometries appear along the wall bisector, where the anisotropy of the stresses in the near-wall region is lower in the equilateral triangular duct.

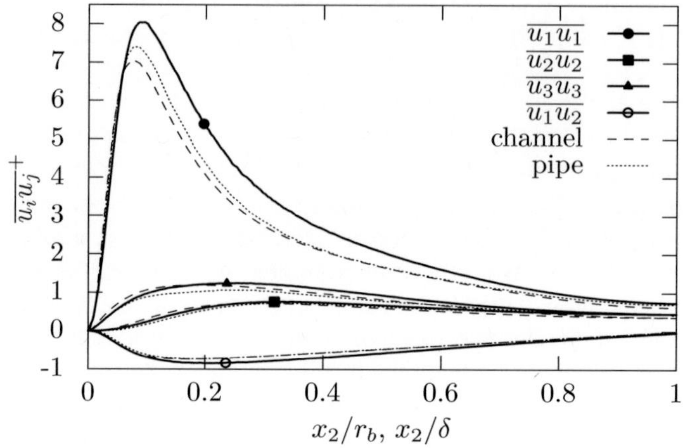

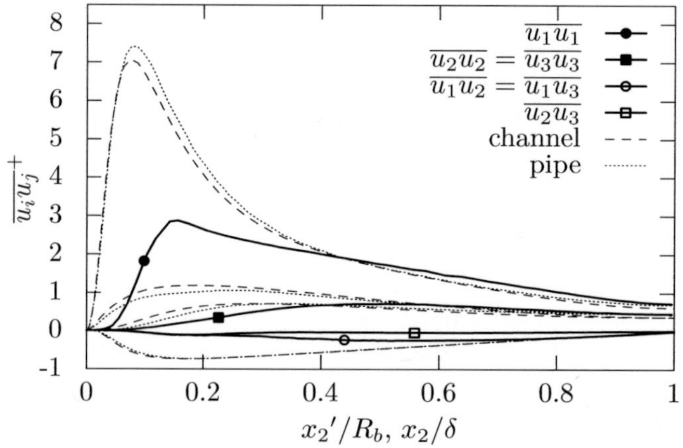

Figure 9.3: Reynolds stresses along the wall bisector (upper plot) and along the corner bisector (lower plot) of the square duct. The data along the corner bisector are plotted starting in the corner and using the local coordinate x_2' along the duct diagonal. u_τ resulting from the simulation that leads to $Re_\tau = 150$ is used for normalization. Literature data for channel flow at $Re_\tau = 180$ [5] and for pipe flow at $Re_\tau = 180$ [32] are shown for comparison and δ is equal to the channel half height and the pipe radius.

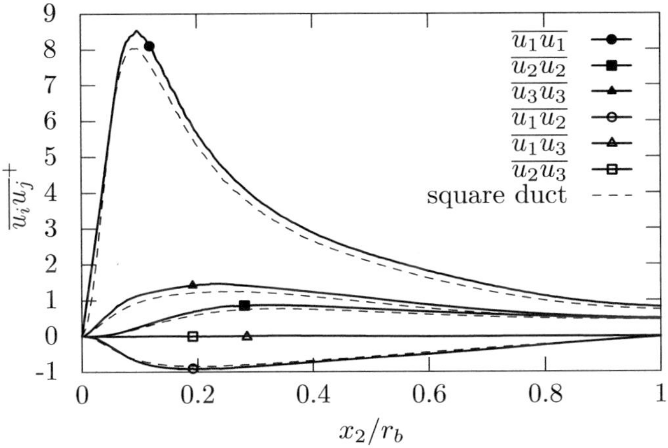

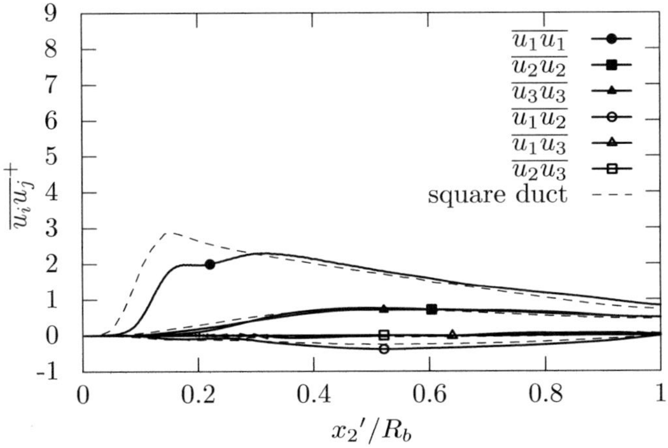

Figure 9.4: Reynolds stresses along the wall bisector (upper plot) and along the corner bisector (lower plot) of the equilateral triangular duct where r_b and R_b is the length of the bisector, respectively. The data along the corner bisector are plotted starting in the corner and using the local coordinate x_2' in negative x_2-direction. The data for the square duct are also shown for comparison. u_τ resulting from the corresponding simulations is used for normalization.

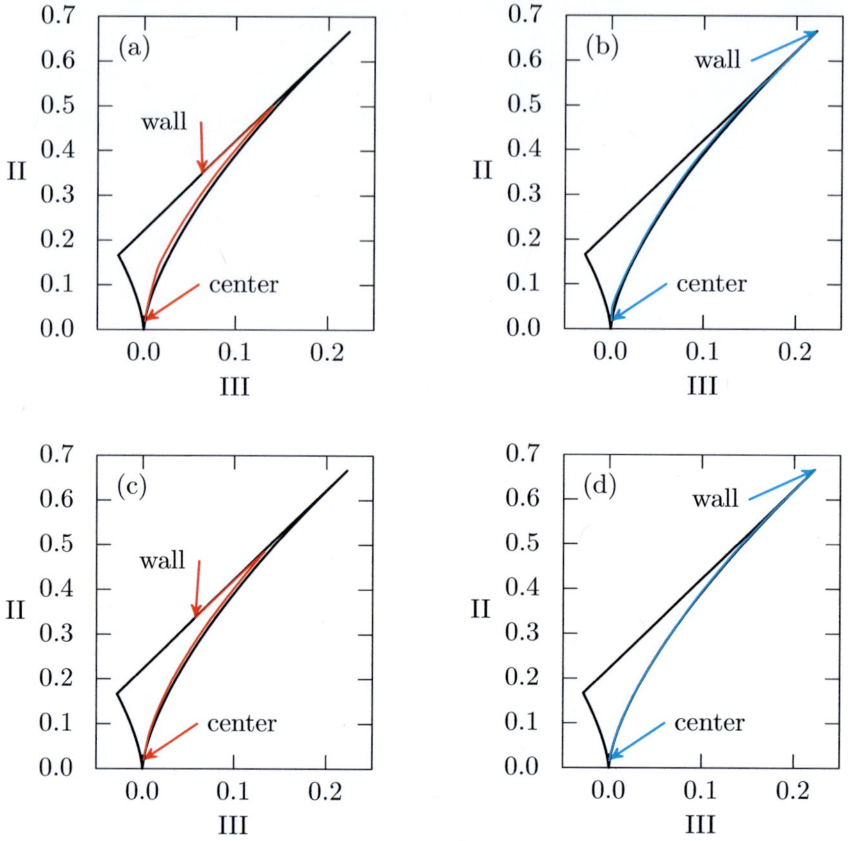

Figure 9.5: Trajectory through the anisotropy invariant map along the wall and corner bisectors of the square duct (a), (b), and the equilateral triangular duct (c), (d). Left column and red color, wall bisector; right column and blue color, corner bisector.

From the observations made for the statistical flow field in the square and equilateral triangular duct, it can be concluded that duct corners with different corner angles lead locally to a strong tendency towards axisymmetry in the stresses. This behavior is observed along the corner bisectors of the ducts. In the vicinity of the corner angle, the disappearance of turbulent properties is observed while the extension of the effect is increased for the smaller angle. In contrast, along the wall bisector, the anisotropy of the stresses and their tendency towards axisymmetry is reduced and the flow shows similar properties to the turbulent channel or pipe flow.
The goal in the following sections is to extend the benefits resulting from duct corners to a larger part of the flow domain. For this purpose, novel duct geometries are designed, which are presented and analysed.

9.2 Design of novel duct geometries using corner effects

In addition to the frequently studied duct flows discussed in the previous section, a few more investigations of turbulent flow through more complex non-circular ducts are presented in the literature. Raiesi *et al.* [102] and Fukushima and Kasagi [43] studied the flow in ducts with rhombic cross sections and focused on the impact of different corner angles on the flow. The smallest angle considered in these investigations is 30° and a tendency towards flow laminarization, similar to the observations made in the previous chapter for the triangular ducts with 4° and 11.5° apex angles, is observed.
Lammers *et al.* [78] performed direct numerical simulations of the turbulent flow through ducts of polygon-shaped cross sections using a Lattice–Boltzmann method. Starting from turbulent flow through the square duct the intention in their work is to increase the anisotropy of turbulence along the entire wetted perimeter due to an increase in corner bisectors. For this purpose, ducts consisting of eight corner regions distributed regularly along the circumference are designed. Within the resulting octagonal cross section, the cases of straight and profiled sides are distinguished, which intersect at corner angles of 135° and 90°, respectively. Along the profiled sides of the duct the anisotropy at the wall is found to increase and reaches almost the one-component limit at the corners. In contrast, for the octagonal cross section with straight sides, no noticeable increase in anisotropy is observed.
The present investigation follows the ideas of Lammers *et al.* In general, two major features of duct geometries are expected to be connected to the statistical behavior of the flow, namely the number of corner bisectors and the design of the corner regions. This work concentrates on analysing the latter factor of relevance. For this purpose, the number of corner bisectors is kept constant and set to nine. This choice is based on experimental findings from Schiller [105], which will be discussed in the context of the present results in Section 9.2.2. Variations in the design of the corner region can generally imply different corner angles and alternatives for the design of the connecting side walls, which are referred to as crests. In this approach, an initial geometry with a star-shaped cross section, which is shown in Figure 9.6 (a), is defined. The corner angle is set to 88°, which is similar to Lammers' promising case. Starting from this cross section, further geometries are derived in Figure 9.6 (b)–(d) by variations of the angle of the duct corners together with the shape of the sides: (b) the crest and the tip regions are rounded, (c) the crest region is rounded while the corner angle is kept constant, (d) the crest region is rounded and the height of the surface corrugation is kept constant.

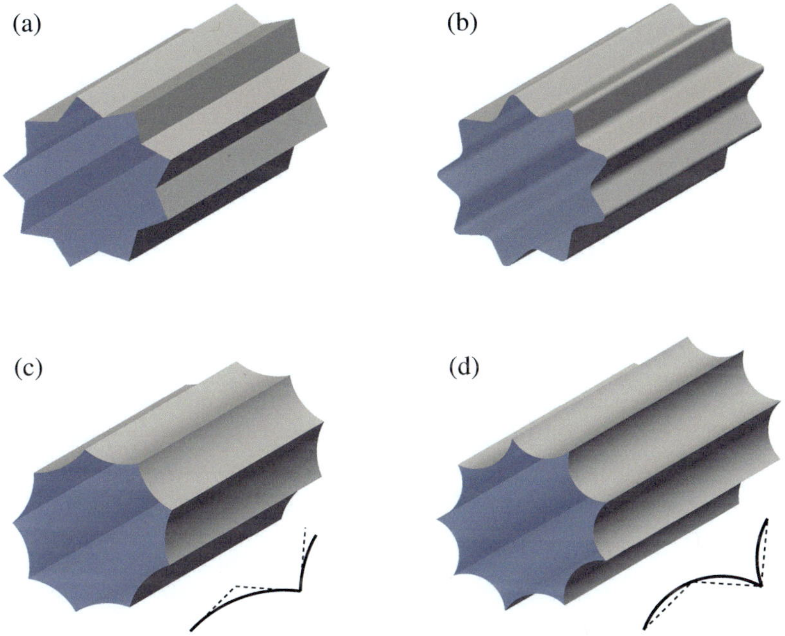

Figure 9.6: Novel duct geometries with differently designed tip and crest regions. (a) Initial geometry, (b)–(d) derived geometries. Case (a) includes edged corner and crest regions, which are rounded in case (b). Additional sketches illustrate similarities and differences in the design of the corner regions in cases (c) and (d) compared with case (a).

The corner and crest design in geometry (c) is fairly similar to Lammers' case but the number of bisectors is changed from eight to nine.

9.2.1 Impact on the turbulent stresses

The statistical flow fields that arise in the duct geometries shown in Figure 9.6 are now investigated. In order to analyse the impact of the differently designed corners and crests on the flow, the data are again plotted along the corresponding bisectors, which are illustrated as an example for geometry (a) in Figure 9.7. In the following, the corner bisector is also referred to as tip bisector.
The Reynolds stresses developing along these lines in case (a) are shown in Figure 9.8. In the plots, the corresponding data for square duct flow (see Figure 9.3)

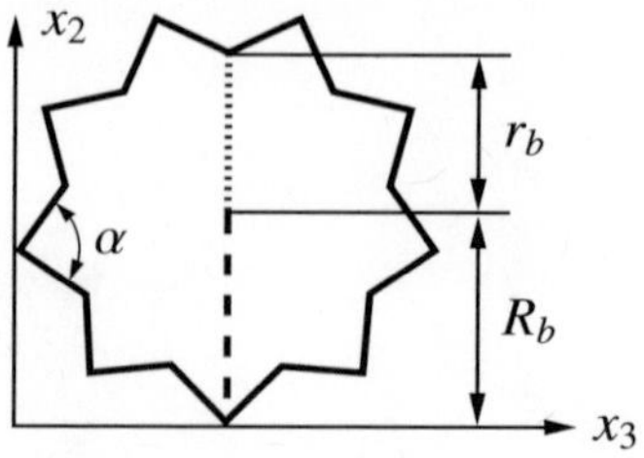

Figure 9.7: Cross section plane of the star pipe with crest and corner bisector. The crest bisector with the length r_b is sketched with a dotted line, the tip bisector with the length R_b is drawn with a dashed line, $\alpha = 88°$.

are also shown. The statistical properties of the flow in the ducts show similarities: the distribution of the stresses along the crest or wall bisector resemble the characteristic for turbulent wall-bounded flows, while the turbulent fluctuations along the tip or corner bisector are damped. Above the crest region of the star-shaped duct the extrema of all stress components are slightly increased and move closer to the wall compared with the mid-wall region of the square duct. The increase is more pronounced for the spanwise component than for the streamwise and wall-normal components. Choi *et al.* [21] observed a similar behavior above the crest of riblets. In general, the effect of the crest on the flow is felt to be somewhat small considering the fact that an acute corner extends into the flow domain. The development of the Reynolds stresses along the tip or corner bisector are shown in the lower plot in Figure 9.8 where the extension of the corner region in the x_2-direction is marked by a vertical line. This height of the corner region corresponds to approximately 50 wall units. The entire suppression of turbulence in the close vicinity of the corners is observed in both geometries. Moving further along the diagonal of the square duct, the geometrical properties of the duct lead to $\overline{u_2u_2} = \overline{u_3u_3}$. This behavior is not observed for the star-shaped pipe. Additionally, the damping of the streamwise stress component, $\overline{u_1u_1}$, is more pronounced than in the square duct. These differences are noticeable due to the fact that locally the geometrical boundaries of the domains are almost identical. Moreover, despite of the fact that the extension of the corner region is limited, the turbulent motion above the tip is affected in large parts of the corner bisector.
In the following discussions, the statistical flow field in the star pipe (case (a)) is compared with the geometries (b)–(d). The extension of the corner region is again marked by a vertical line in the development of the Reynolds stresses above the tip. The height of the corner region is almost identical in the geometries (a), (b) and (d) and only smaller in geometry (c) (compare the sketches in Figure 9.6). The

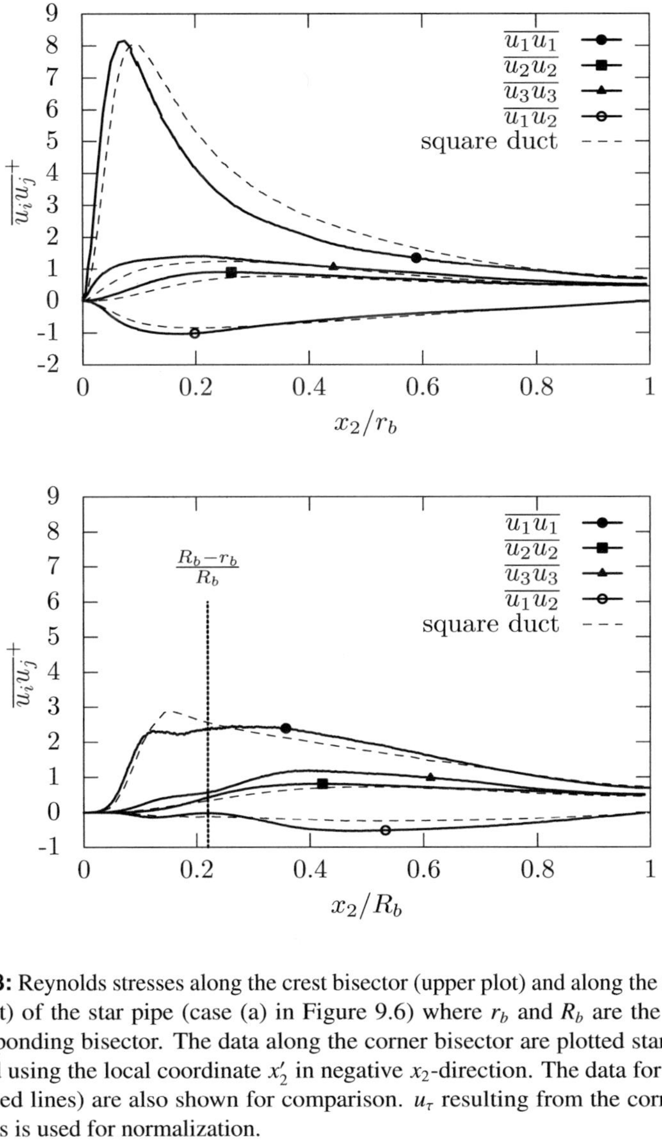

Figure 9.8: Reynolds stresses along the crest bisector (upper plot) and along the tip bisector (lower plot) of the star pipe (case (a) in Figure 9.6) where r_b and R_b are the lengths of the corresponding bisector. The data along the corner bisector are plotted starting in the corner and using the local coordinate x_2' in negative x_2-direction. The data for the square duct (dashed lines) are also shown for comparison. u_τ resulting from the corresponding simulations is used for normalization.

data arising from the different duct geometries are generally fairly similar. The discussion will focus on highlighting the differences that appear.

In Figure 9.9 the effect of rounded crest and tip regions is illustrated. The effects are minor and only a reduced damping of $\overline{u_1u_1}$ when moving away from the tip is observed.

The effect of entirely curved walls on the development of the stresses is shown in Figure 9.10. The smoothing of the crest region, which also is accompanied by the reduced extension in the flow compared with the other geometries investigated, results in a decrease in the magnitude of all stress components in the near-wall region. At the same time, the stresses in the vicinity of the corner are damped in a similar manner as observed for the star-shaped duct. However, in particular the $\overline{u_1u_1}$-component increases faster in a certain distance from the tip of duct (c) compared with the star-shaped pipe. This fact could be associated with the spatially restricted influence of the corner region.

Finally, the impact of a stronger curvature of the side walls that is associated with a smaller corner angle compared with the previous cases is analysed based on Figure 9.11. In contrast to the observations made for geometry (c), a rather high peak value of $\overline{u_1u_1}$ above the crest region shows up. At the same time, the development of the Reynolds stresses along the corner bisector highlights an increased region, in which turbulent fluctuations are strongly damped. Contrary to the duct geometries investigated previously, entire laminarization of the flow is observed close to the acute corner. Moving away from the corner, the stress component $\overline{u_1u_1}$ develops similar to geometries (a) and (b) and a flattened peak value is observed. This similarity together with the findings for geometry (c) suggest that the depth of the surface structure rather than the corner angle is responsible for this property of the flow.

In summary, the development of the stresses in the ducts composed of nine corner regions indicate a slight enhancement of turbulence above the crest regions and a significant damping in the corner regions. It is of particular interest in the context of this work that the beneficial effect of the corner region extends far into the flow field in order to compensate the expected negative effects originating from the crests.

In order to gain further insights into the configuration of the Reynolds stresses in geometries (a)–(d), the development of the anisotropy tensor is plotted in invariant space. In Figure 9.12 the data are shown for geometries (a) and (b). While the behavior is fairly similar in the center of the ducts, differences can be observed at the wall. The anisotropy above the rounded crest of geometry (b) is increased compared with geometry (a). The rounded corner region leads to the opposite effect and the one-component state is not reached.

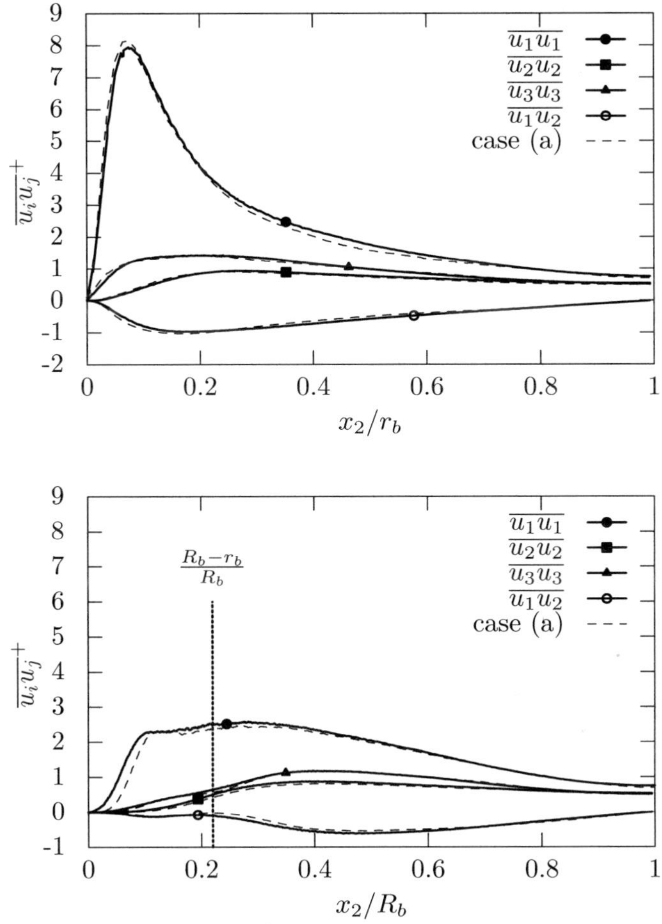

Figure 9.9: Reynolds stresses along the crest bisector (upper plot) and along the tip bisector (lower plot) of case (b) in Figure 9.6 where r_b and R_b are the lengths of the corresponding bisectors. The data along the corner bisector are plotted starting in the corner and using the local coordinate x_2' in the negative x_2-direction. The data for the pipe with star-shaped cross section (dashed lines) are also shown for comparison. u_τ resulting from the corresponding simulations is used for normalization.

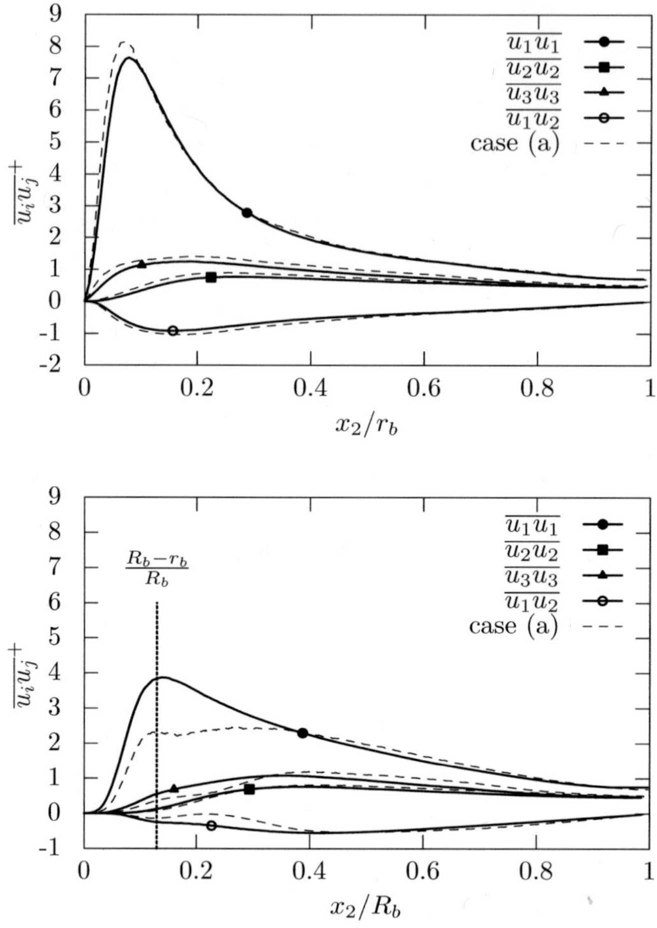

Figure 9.10: Reynolds stresses along the crest bisector (upper plot) and along the tip bisector (lower plot) of case (c) in Figure 9.6, where r_b and R_b are the lengths of the corresponding bisectors. The data along the corner bisector are plotted starting in the corner and using the local coordinate x_2' in the negative x_2-direction. The data for the pipe with star-shaped cross section (dashed lines) are also sketched for comparison. u_τ resulting from the corresponding simulations is used for normalization.

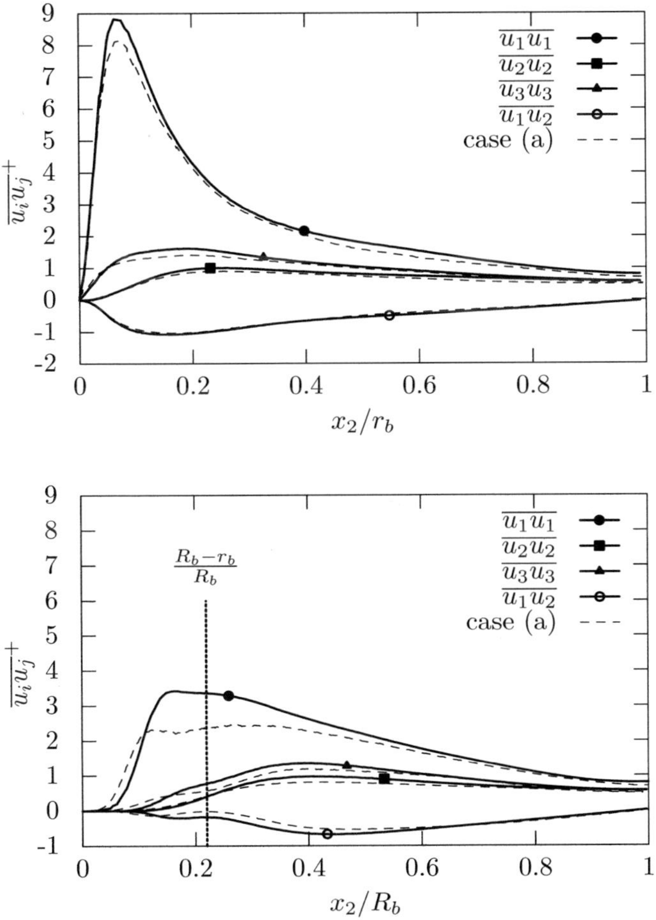

Figure 9.11: Reynolds stresses along the crest bisector (upper plot) and along the tip bisector (lower plot) of case (d) in Figure 9.6 where r_b and R_b are the lengths of the corresponding bisectors. The data along the corner bisector are plotted starting in the corner and using the local coordinate x_2' in the negative x_2-direction. The data for the pipe with star-shaped cross section (dashed lines) are also shown for comparison. u_τ resulting from the corresponding simulations is used for normalization.

Above the corner regions of both ducts, a general trend in the data to approach the axisymmetrical border of the map is seen.
The development of the invariants of the anisotropy tensor for geometries (c) and (d) are presented in Figure 9.13. Again no significant difference can be observed in the center of the ducts. The same holds for the behavior above the crest and in the corner where in both ducts similar anisotropy levels are reached. For both geometries, the one-component state is reached within the corner edge and a tendency towards axisymmetry along the corner bisector is present.
In summary, the flow properties along the corner bisectors coincide with the theoretically derived constraints for low energy dissipation in the flow. However, the beneficial impact of the corners does not persist towards the crest regions. In the following paragraph, an attempt is made to link the present local observations in the stress field to global trends in the dissipative losses. From this analysis, the potential of different corner shapes for the present goals, suppressing turbulent dissipation and delaying the transition to turbulence, is assessed.

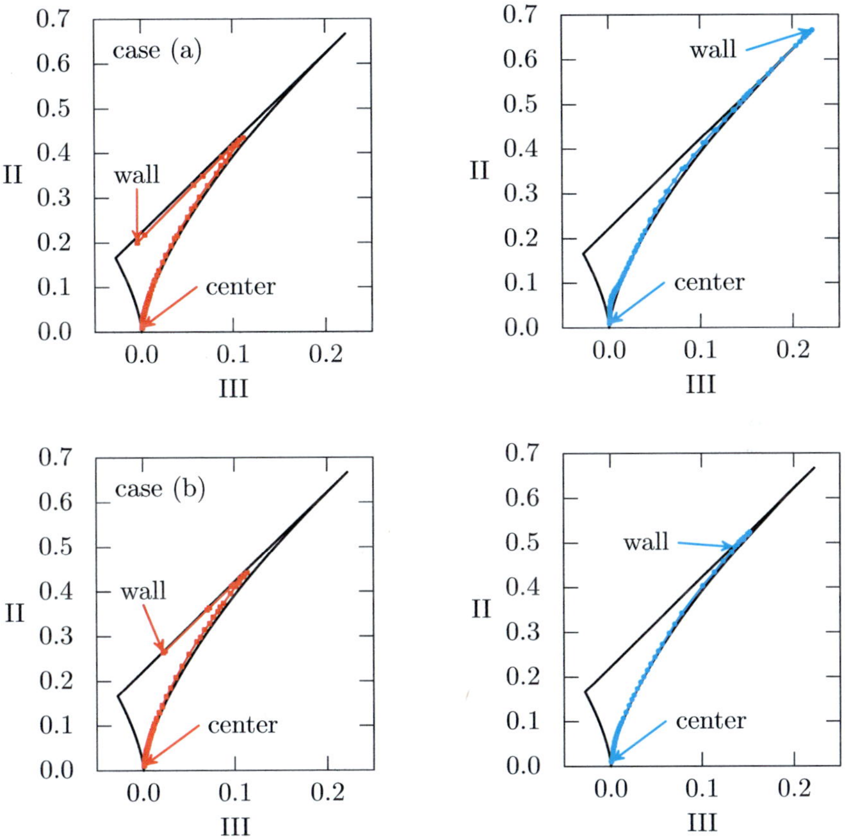

Figure 9.12: Trajectory through the anisotropy invariant map along the crest and tip bisectors of cases (a) and (b). Left column and red color, crest bisector; right column and blue color, tip bisector.

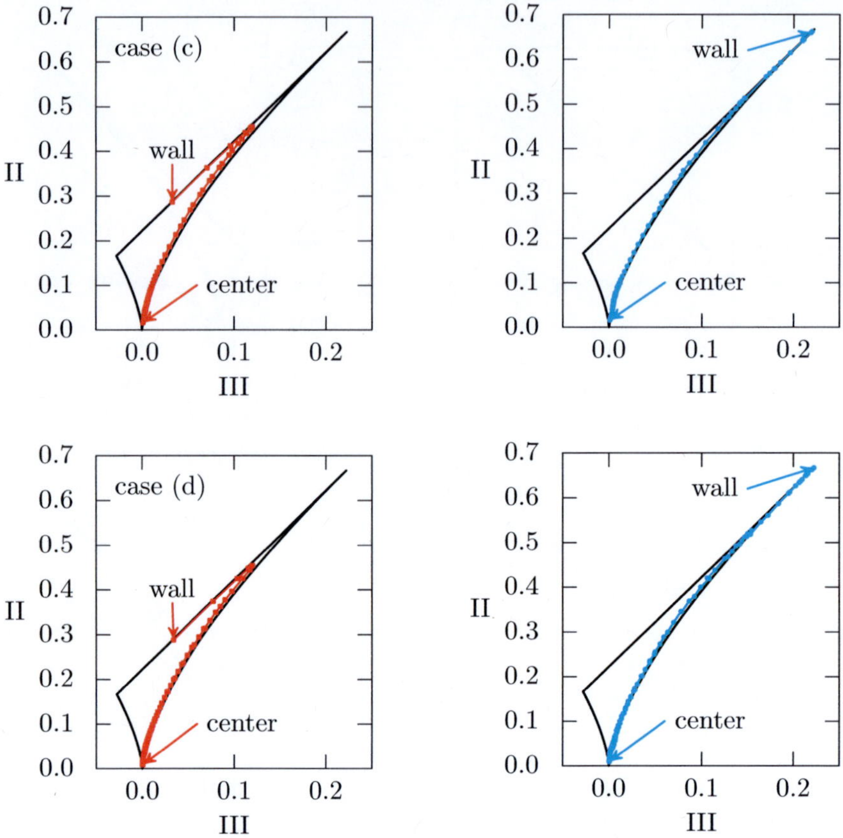

Figure 9.13: Trajectory through the anisotropy invariant map along the crest and tip bisectors of cases (c) and (d). Left column and red color, crest bisector; right column and blue color, tip bisector.

9.2.2 Impact on the dissipative losses in the flow

The effect of the observed local properties of the stress distribution on the dissipative losses in the flow is discussed in this section. For this purpose, the present results for the global frictional resistance of geometry (b) are first complemented by experimental results available in the literature. Further, the production of kinetic energy in the cross section and the dissipation along the wall are used for comparative analysis of all duct shapes. The potential of the individual corner designs to interact with the dissipative nature of turbulence is assessed.

Schiller [105] investigated the flow resistance of geometry (b) experimentally up to $Re_h = 40000$. The experimental data together with the results from the simulations in the present work are shown in Figure 9.14 where a satisfactory agreement of the data is observed [25]. Besides the numerical result for $Re_h = 4500$, an additional simulation is performed at $Re_h = 10000$. At the lower Reynolds number, no significant deviation from the Blasius correlation is observed. Interestingly, a least square fit shows that the data tend to progressively lie below the Blasius correlation for increasing Reynolds numbers. At the same time, the entire dissipation of the flow is known to be progressively dominated by turbulent dissipation if the Reynolds number increases [76]. Thus, the observed decrease of the flow resistance might be brought about by a reduction in the turbulent dissipation. In the context of the previously discussed localized effects on the configuration of the stresses, the data for the friction factor suggest that overall the beneficial effects in corner regions prevail over the critical behavior above the crest.

A data set of the friction factor arising in the duct geometries (a), (c) and (d) at increasing Reynolds numbers is not available. However, since the observations made for the stress fields in the different ducts in general are similar to geometry (b), similar implications on the trend in the overall turbulent dissipation are also expected.

Further investigations follow the physical model summarized in Figure 6.4. These suggest, that reduced energy dissipation in turbulent flow is strongly related to low production of kinetic energy of turbulence in the entire domain and low turbulent dissipation, ϵ, at the wall and that these states are linked to high axisymmetry in the stresses.

The trajectories in the anisotropy maps in Figures 9.12 and 9.13 have shown a strong trend towards axisymmetry along the corner bisectors. This behavior is assumed to be associated with a reduced production of kinetic energy of turbulence at these positions. In Figure 9.16 the contour plot of P_k arising in geometry (b) is shown where P_k is evaluated according to Equation (6.11). For comparison, a contour plot of P_k arising in a cylindrical pipe at the same hydraulic Reynolds number, $Re_h = 4500$, is also presented.

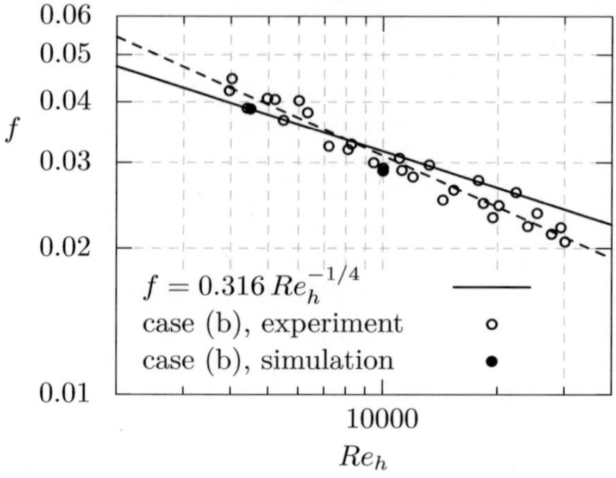

Figure 9.14: Friction factor f plotted over the hydraulic Reynolds number Re_h for geometry (b). The experimental data from Schiller [105] are shown together with data from present DNS studies elaborated in collaboration with Krieger [72]. A least squares fit of the data (dashed line) and the Blasius correlation [106] (solid line) are included.

In this configuration, $P_k \approx 80$ is the maximum that is reached in the wall region of the pipe. This value, as well as the distribution of P_k, is in good agreement with findings from Eggels *et al.* [32] for circular pipe flow at a similar Reynolds number. It can be seen, that the production of kinetic energy of turbulence is lower within the corner regions of duct geometry (b) than along the wall of the cylindrical pipe. Additionally, the magnitude of P_k remains low when moving from the corner further towards the duct center. Since the color scale used for the ducts is the same, Figure 9.15 also highlights that P_k above the crest regions of geometry (b) is increased compared with the maximum appearing in the cylindrical pipe. Clearly, this trend contrasts with the beneficial effect observed in the corner region.
In Figure 9.16, the distribution of P_k in all of the novel designed ducts is considered. In general, the impression is similar for all shapes: The production of turbulent kinetic energy is high above the crest region but this behavior marginally affects the flow in the proximity of the corner bisectors. There, the self-sustaining mechanism of turbulence can be assumed to be strongly damped. The chosen color scales also demonstrate the maximum values of P_k for the individual ducts. In this comparison, the maximum values of P_k above the crests are found to be the smallest for duct geometry (c).

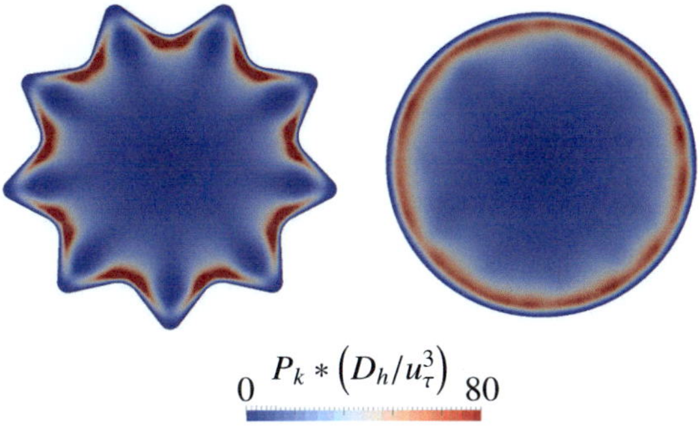

Figure 9.15: Production of kinetic energy of turbulence, P_k, in the cross section of duct (b) and a cylindrical pipe. Both calculations are run at $Re_h = 4500$ and u_τ resulting from the corresponding simulation is used for normalization.

In Section 6.2, reaching the one-component state of the fluctuations, and thus maximum anisotropy at the wall, was linked to the entire suppression of dissipation originating from the fluctuating motion [66]. Thus, an increase in the anisotropy at the wall was suggested to be correlated with reduced turbulent dissipation. This trend is confirmed by literature data for wall-bounded flows in Figure 6.2. The corresponding illustration is reconsidered and applied to the interpretation of the results for non-circular duct flows. For this purpose, the first available data arising at the corner and wall bisector of the square duct flow at $Re_h = 10300$ [58] are included, which agree fairly well with the extrapolated trend in the plot. These data correspond to the extrema arising along the duct side and highlight that the dissipation in the corner, where the stresses reach the one-component state, is significantly decreased compared with the wall bisector.
The extrema in the anisotropy level that are reached on the crest (red color) and in the corner (blue color) of the non-circular ducts studied in this section are also indicated in the plot. For all geometries with acute corner edges, the one-component state of the stresses is achieved in the corner. These geometries have corner angles of 88° and 42°. For a corner angle of 135° Lammers *et al.* [78] did not observe this high level of anisotropy. Thus, the question still remains as to which angle limits the possibility of reaching the one-component state in the corner.
Figure 9.17 also shows that $(\mathrm{II})_{\mathrm{wall}}$ is significantly smaller above the crests. It is noted that the magnitude of anisotropy at these positions is also decreased

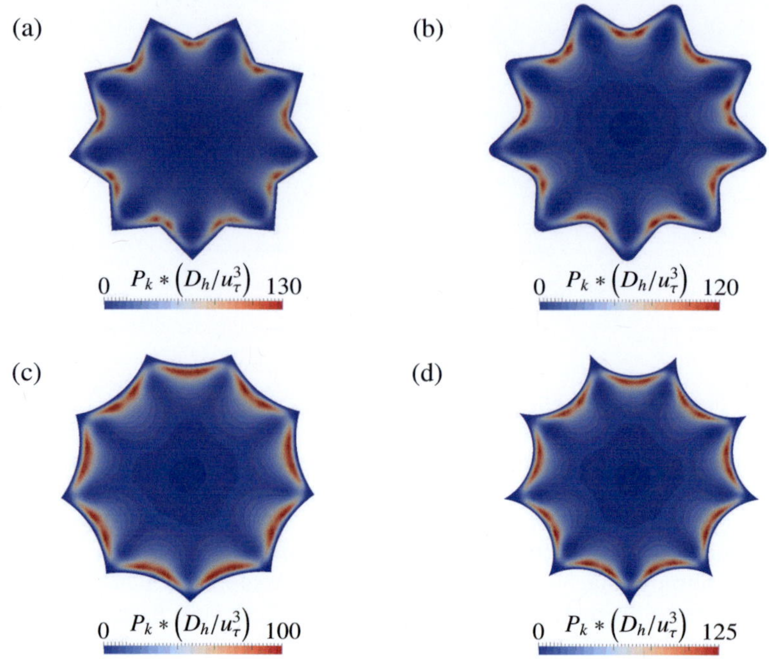

Figure 9.16: Production of kinetic energy of turbulence, P_k, in the duct cross sections (a)–(d). u_τ resulting from the corresponding simulation is used for normalization.

compared with the value evaluated for circular pipe flow at a similar Reynolds number, where $(\mathrm{II})_{\mathrm{wall}} = 0.37$ [32]. In addition, the comparison of $(\mathrm{II})_{\mathrm{wall}}$ reached above the crests of the duct geometries (a)–(d) to the value arising for the wall bisector of square duct flow can provide insights into the impact of the wall curvature on the dissipation. Generally, ϵ^+ depends on the Reynolds number. Thus, the anisotropy level that is reached on the wall bisector of square duct flow at a similar Reynolds number is also indicated in the plot (green line). This suggests that the dissipation above the crests of the ducts (a)–(d) is significantly higher than above the flat wall of the square duct. Comparing the minima in $(\mathrm{II})_{\mathrm{wall}}$ arising at the crests, the lowest value is found for geometry (a) implying highest turbulent dissipation. In contrast to the remaining geometries, the crest is not rounded in this case.

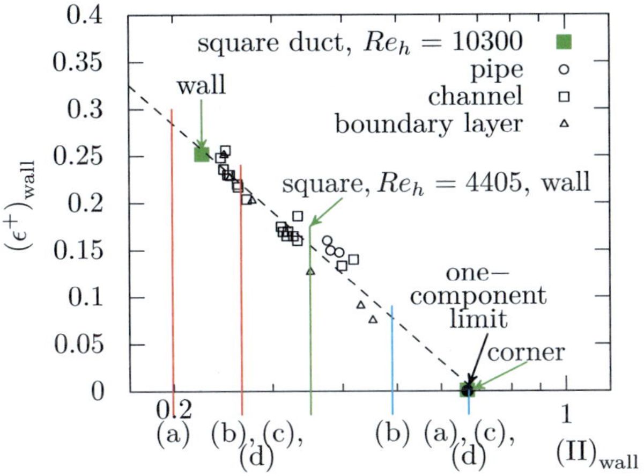

Figure 9.17: Extension and application of Jovanović and Hillerbrand's [67] illustration of the turbulent dissipation rate at the wall *versus* the magnitude of anisotropy at the wall for analyses of non-circular duct flows. The references for numerical data of (cylindrical) pipe, channel and boundary layer flows are given in Figure 6.2. Additionally, the data for the corner and wall bisector of square duct flow at Re_h = 10300 from Huser and Biringen [58] are plotted and the anisotropy level on the wall at Re_h = 4405 is shown. The anisotropy levels at the wall arising from the present novel duct geometries (a), (b), (c) and (d) are also included: red color, crest bisector; blue color, corner bisector (compare also Figures 9.12 and 9.13).

The comparison of the extrema in the anisotropy levels reached at the duct walls suggests that overall $(II)_{wall}$ is highest for the geometries with acute corner angles and profiled sides, (c) and (d) and that this configuration of the stresses is associated with the smallest turbulent dissipation. This connection approves these geometries for the present attempt towards the design of duct geometries leading to the laminarization of turbulent flow. Since the one-componentality in the apparent stresses at the wall is also found to provoke the persistence of laminar flow at high Reynolds numbers, these geometries are also suggested to have largest potential for the delay of transition.

Previous analyses have considered the production of kinetic energy in the cross section plane and the distribution of the turbulent dissipation at the wall. These assessment criteria suggest that geometry (c) has the highest potential for the present purposes, namely the reduction of turbulent dissipation and the delay of laminar to turbulent transition. However, in general the impact of differently shaped corner and crest regions is similar and the flow is only affected locally in the required manner. In the most beneficial case, turbulent dissipation would vanish across the entire flow field leading to laminar flow, which is not observed in any case. The numerical simulation of the laminarization process was not intended in the present approach (see Figure 6.7), since it results in a large numerical effort for the complex duct geometries being considered. However, it is felt that this fact requires to be pointed out. It is discussed further in the next paragraph.

9.2.3 Destabilization of the mean velocity profile

Limits for flow laminarization

The results in the previous paragraph indicate that turbulent dissipation is locally reduced in duct geometries, including corner regions. This behavior is initiated by beneficial effects on the statistical properties of the flow in the proximity of the corner bisectors that locally lead to flow states that are related to laminar flow. However, since the flow is not entirely laminarized, an attempt is made to consider restrictions for the process. The discussion is based on the instabilities induced by inflectional velocity profiles that can be connected to the appearance of turbulence sustaining dynamic processes in the flow. Velocity profiles appearing in geometry (b) are studied in detail in this context.

The velocity field of turbulent wall-bounded flows is characterized by the instantaneous formation of coherent structures in the form of quasi streamwise vortices [99]. The cross flow associated with these vortices transports slow moving fluid away from the wall leading to a region of reduced streamwise velocity, which is typically referred to as low speed streak. The streaks have a characteristic behavior, known as bursting. With increasing downstream distance, a streak migrates away from the wall, which leads to the ejection of low-speed fluid in the bulk flow. This process is found to be largely responsible for the maintenance of turbulence in wall-bounded flows. At the same time, a vortex transports higher-speed fluid from the bulk down towards the wall leading to so-called sweeps.

The appearance and dynamics of coherent structures in turbulent wall-bounded flows have been studied extensively in the past. Numerous of these investigations deal with the interaction of the coherent structures with the stability of the flow. In this context it was pointed out by Blackwelder [13] that the instantaneous streamwise velocity profile over flat walls is typically characterized by inflection points. These inflection profiles are generally found in the wall-normal as well as in spanwise direction [14]. Further, in experimental investigations Kline *et al.* [70]

observe that strong inflection profiles are a characteristic feature of the bursting of streaks, which represents a prominent instability mechanism.

Holmes *et al.* [56] consider the particular appearance of the instantaneous streamwise velocity profile that is formed in connection with a streak. This profile is expected to have an inflection point in the (x_1, x_2)-plane and two such points in the (x_1, x_3)-plane, where x_1 corresponds to the streamwise, x_2 to the wall-normal and x_3 to the spanwise direction. The authors argue, that either the latter pair of inflection points, or the former one, could be responsible for the instability, since an inflectionary profile is inherently unstable. It is proposed that the growth rate of the instability associated with the inflection points is dependent on the shear rate at these points, $\partial U_1/\partial x_2$ and $\partial U_1/\partial x_3$.

In turbulent flow over a flat wall, the shear in the spanwise direction vanishes in the temporally averaged velocity statistics. However, the instantaneous appearance of coherent structures in the flow permanently induces shear and instability in the flow in an instationary fashion. Surface structures, which are aligned with the mean flow direction, additionally lead to variations of the temporally averaged streamwise velocity profile in the spanwise direction. Thus, the presence of these structures can be believed to permanently destabilize the flow. Despite the instability induced in the spanwise direction, the surface structures might also affect the velocity profile in the wall-normal direction.

These ideas on the impact of surface structures on the mean velocity profile can be transferred to the duct shapes studied previously. To do this, the corresponding flow field is chosen to be compared with the flow properties in a circular pipe in a cylindrical frame of reference. This coordinate system for a segment of geometry (b) is indicated in Figure 9.18 together with variables that are used in the following discussion.

In turbulent flow through a circular pipe, the mean streamwise velocity, $\overline{U}_z$, is a function of the radial position and does not vary in the circumferential direction. A different behavior is expected within the profiled duct shapes. The left plot in Figure 9.19 shows the turbulent mean streamwise velocity profile along circular arcs at different radial positions in geometry (b). The symbols for these positions are chosen according to Figure 9.18 (left). When moving in the positive radial direction, it can be observed that circumferential variations in the streamwise velocity strongly increase when the duct wall is approached. At the position $r/r_b = 0.9$ the variation is fairly small but becomes significant at $r/r_b = 0.99$. The strong variation is mainly due to the fast growth of the velocity above the valley region ($\varphi^+ = 0$) were similar values of $\overline{U}_z$ are reached at the considered radial distances. The observed variation of the velocity in circumferential direction leads to inflectional points in the profiles.

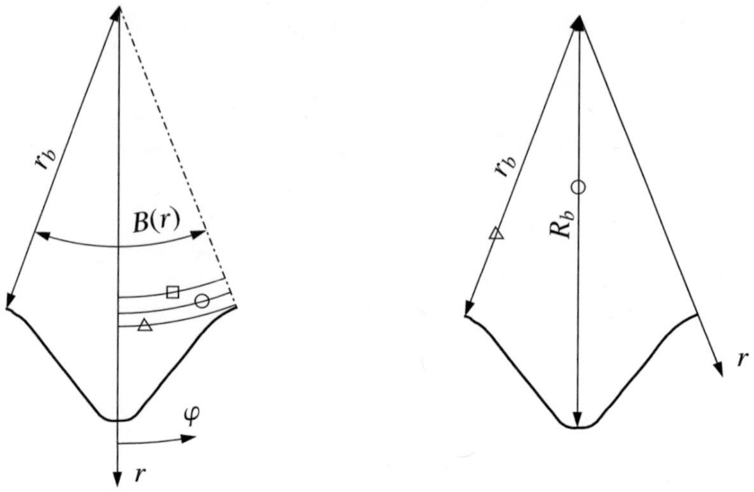

Figure 9.18: Segment of the cross section plane of geometry (b) with a cylindrical coordinate system. Left sketch: investigated trajectories in circumferential direction together with allocated symbols. Circular arcs at the positions $r/r_b = 0.9$ (square symbol), $r/r_b = 0.95$ (circular symbol) and $r/r_b = 0.99$ (triangular symbol) are shown. $B(r)$ denotes the length of a circular arc. Right sketch: investigated trajectories in the radial direction. Allocated symbols for the valley (circular symbol) and crest bisector (triangular symbol) with the length R_b and r_b, respectively.

Before these points are considered in more detail, the development of $\overline{U_z}$ in radial direction is also studied. For this purpose, the profiles along the corner and crest bisector are compared in the right plot of Figure 9.19. Additionally, the velocity profile arising in circular pipe flow at a similar Reynolds number is sketched [32]. The symbols which are used in the plot are introduced in Figure 9.18 (right). It can be seen, that the deviation of the mean streamwise velocity profile along the crest bisector from the profile arising in circular pipe flow is marginal. Thus, it is concluded that no additional instability is introduced due to the modified duct shape at this position. By contrast, the development of $\overline{U_z}$ along the corner bisector differs significantly from the behavior above the crest and inflection points can clearly be identified in the profile.
However, the previously observed variations in the Reynolds stresses in the proximity of the corner bisector towards statistical axisymmetry suggest a limited effect of the inflection velocity profile on the stability of the flow. In the statistical frame, the relationship of gradients in the mean flow and the Reynolds stresses determines

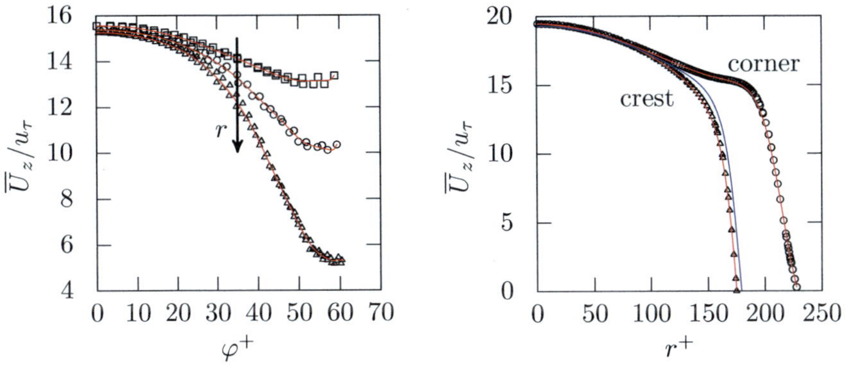

Figure 9.19: Streamwise mean velocity profiles, $\overline{U}_z$, along circular arcs (left plot) and along the bisectors (right plot) for turbulent flow at $Re_h = 4500$ in geometry (b). The trajectories in the flow domain together with the corresponding icon shapes were introduced in Figure 9.18. The data are fitted using bezier curves (red lines). The velocity profile of circular pipe flow at $Re_h = 5300$ is also shown in the right plot (blue line) [32].

the production of the kinetic energy of turbulence. In the cylindrical frame of reference, P_k in the streamwise direction z is defined according to Rotta [104] as

$$P_k = \overline{u_z u_r}\frac{\partial \overline{U}_z}{\partial r} + \overline{u_z u_\varphi}\frac{\partial \overline{U}_z}{\partial \varphi} \quad . \tag{9.1}$$

The distribution of the magnitude of the shear stress components appearing in this equation are shown in Figure 9.20. It is evident that very small values are reached along the corner bisector. Thus, the production of turbulent kinetic energy is low at this position, even if gradients in the mean velocity are present.

Thus, the inflectional points induced by variations of the velocity profile in the circumferential direction are assumed to be more critical for the stability of the present flow situations. According to Holmes *et al.* [56] the destabilizing effect can be further evaluated taking the shear at the inflection points into account. For the flow case considered here, the instability induced at these points is suggested to be associated with a number such as $(\partial\overline{U}_z/\partial\varphi)\,\nu/u_\tau^2$. For the calculation of velocity derivatives in the circumferential direction, the velocity data in Figure 9.22 (left) are approximated by bezier curves. This procedure certainly includes some inaccuracies. However, Figure 9.21 might serve for the discussion of major trends. In general, the inflection point is found between the corner and the crest bisector.

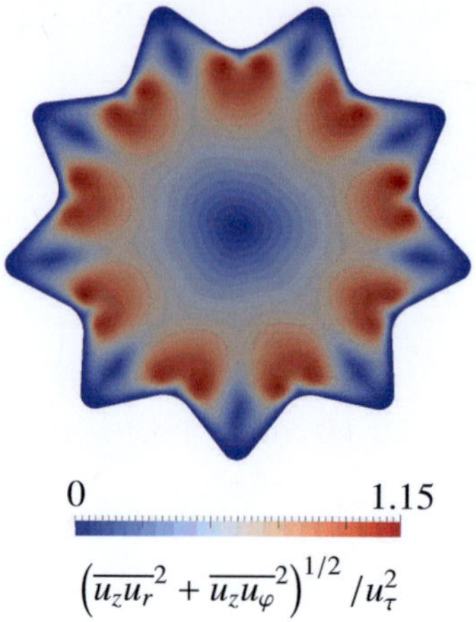

Figure 9.20: Magnitude of the shear stress components $\overline{u_z u_r}$ and $\overline{u_z u_\varphi}$, in the cross section of duct (b). $Re_h = 4500$.

The shear and therefore the instability associated with this point significantly increases when moving in the positive radial direction. For the position closest to the wall ($r/r_b = 0.99$), the inflection point is found fairly close to the crest region of the duct. At this position, statistical axisymmetry in the flow is not preserved and turbulence is assumed to be amplified by the induced instability.
The present attempt towards the identification of restrictive mechanisms for flow laminarization shows that variations of the mean streamwise velocity profile in circumferential direction provoked by the duct shape might be critical in this respect. The findings suggest that the position of inflection instabilities and the associated shear should be considered in future investigations aimed at the preferred design of the duct shapes.

Potential for the persistence of laminar flow

In Chapter 8 it was argued that the same statistical properties are associated with the laminarization of turbulent flow and the persistence of laminar flow. It was shown that corner regions in ducts can provoke the required statistical states in both flow regimes. However, in the previous paragraph it was discussed that inflection instabilities might be a further parameter influencing the process of flow laminarization. Similar indications exist for the persistence of laminar flow. The

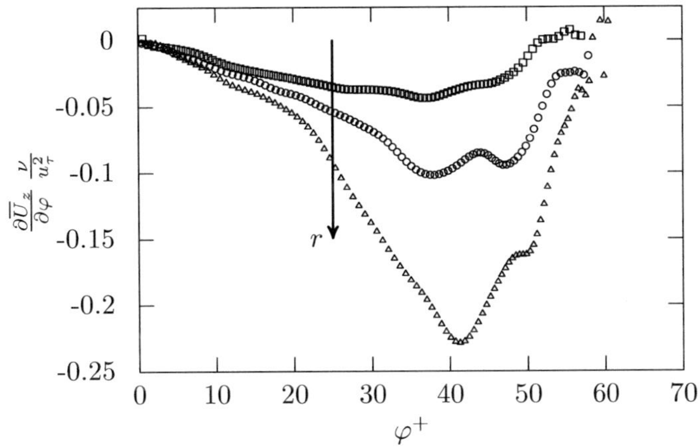

Figure 9.21: Shear of the streamwise velocity profiles $\overline{U}_z$ in the circumferential direction along circular arcs in geometry (b). The derivatives are calculated from the fitted profiles in Figure 9.22 and symbols are used accordingly. $Re_h = 4500$.

impact of velocity profiles with inflection points on the stability of the laminar boundary layer has been frequently studied in the past [106]. Recent investigations by Hof *et al.* [55] suggest that inflection points are also characteristic for the laminar to turbulent transition process in the flow through a cylindrical pipe.
In general, the modified duct shapes are assumed to lead to inflection velocity profiles in the laminar and in the turbulent flow regime. However, since the velocity distribution in the flow regimes is different, differences are also expected in the destabilization of the flow.
Based on the findings for turbulent flow, the variations of the flow in the radial direction are not believed to be critical in this respect and the following analyses concentrate on the development in circumferential direction. The profiles along circular arcs at the radial positions, which were previously analysed in the turbulent regime, are shown in the left plot in Figure 9.22. Owing to the physical importance in the laminar regime, the data are normalized with the bulk velocity U_b. In laminar flow of a given medium that is governed by the Poisson equation (5.4), scaling with U_b provokes the collapse of velocity distributions in a certain duct geometry at different Reynolds numbers, as can be concluded from White's findings for viscous flow [114]. For the normalization of the spatial coordinate, the length of the corresponding circular arc, $B(r)$, is used (see Figure 9.18, left). In order to compare the behaviour in both flow regimes, the mean streamwise velocity field

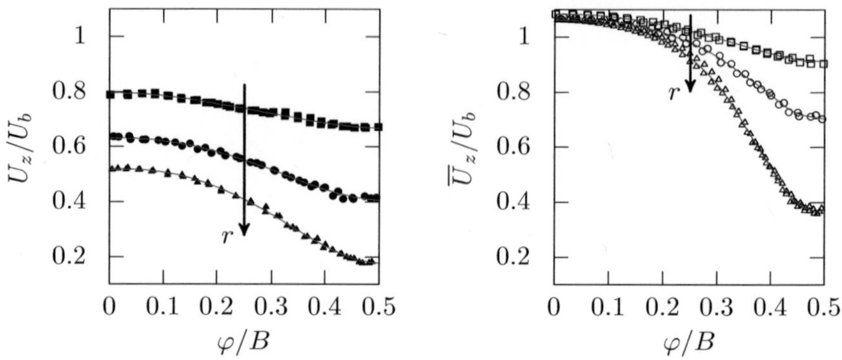

Figure 9.22: Streamwise laminar and turbulent velocity profiles U_z and $\overline{U}_z$ along circular arcs at the positions $r/r_b = 0.9$ (square symbol), $r/r_b = 0.95$ (circular symbol) and $r/r_b = 0.99$ (triangular symbol) in geometry (b). The circular arcs are given in Figure 9.18. Left plot: turbulent flow at $Re_h = 4500$. Right plot: laminar flow at $Re_h = 4500$. The data are fitted (red lines) using polynomial functions and bezier curves, respectively.

arising in turbulent flow at $Re_h = 4500$ in geometry (b) is also normalized with U_b in the right plot of Figure 9.22. Clearly, the development of the flow in circumferential direction is significantly smoother in laminar flow than in turbulence. The difference in the profiles is based on the fact that in laminar flow the velocity distribution is only governed by viscous effects leading to a less rapid increase above the corner ($\varphi/B = 0$). In contrast to the observations made for turbulent flow, the maximum value of U_z is different at all radial position investigated.
The differences in the velocity profiles in both flow regimes also indicate differences in the resulting shear at the inflection points. For the estimation of the shear distribution, the fitted lines through the data in Figure 9.22 are used and the result is shown in Figure 9.23. For the applied scaling, it can be observed that in general the shear is significantly higher in turbulent than in laminar flow.
The previous analyses suggest that the instability in the velocity profile induced by the duct shape is less in laminar flow than in turbulence. Thus, it can be concluded that the delay of transition to turbulence due to appropriate shaping of the ducts generally is easier to establish than the laminarization of turbulent flow. In laminar flow, the potential to prevent disturbances in the flow from increasing can be imagined to prevail over the low instability in the flow profile induced by spanwise variations in the boundary.

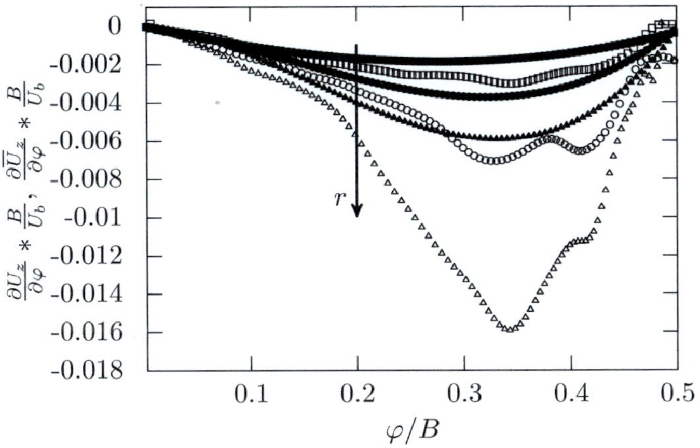

Figure 9.23: Shear of the laminar and turbulent streamwise velocity profiles U_z and $\overline{U}_z$ in the circumferential direction in geometry (b). The data are plotted along the circular arcs given in Figure 9.18. The derivatives are calculated from the fitted profiles in Figure 9.22 and symbols are used accordingly. $Re_h = 4500$.

The conclusions made for the destabilization of the flow due to the circumferential shear rates of the turbulent and laminar flow in the duct can also be discussed in the context of previous findings for surface structures aligned with the mean flow. In general, it can be expected that surface structures are more likely to lead to the delay of transition to turbulence than to laminarization of turbulent flow. This conclusion is in agreement with experimental findings from Jovanović *et al.* [64] for surface embedded grooves.

10 Quantification of potential energy savings

This work aims at providing solutions for the optimization task formulated in Section 4.4. The reduction of the energy dissipation associated with a given flow rate is intended to be induced by changes in the cross section shape of internal flow domains. Equation (4.10) suggests that the pumping power can be used to quantify possible energy savings, which are of interest in this context.
In Chapter 5, losses appearing in laminar flow were discussed. Since the pressure difference along the length of the flow domain, $\Delta P/l$, resulting from a given volume flow rate, $\dot{V}$, is analytically shown to be minimal in a pipe with a circular cross section, the same conclusion follows for the pumping power [107].
At higher Reynolds numbers, for which the flow changes to the turbulent state, the situation is more complex: the circular pipe can no longer be proven to be optimal and corner regions in ducts are suggested to be a beneficial influence on the flow. In Section 9.2, the impact of various corner designs on the physical properties of the turbulent velocity field at $Re_h = 4500$ were investigated. These flows will now be reconsidered under the present energetic aspect. The geometries are labeled as given in Figure 9.6, namely as (a)–(d).
In Figure 9.14, experimental and numerical values for the friction factor, f, arising in duct (b) are plotted together with the estimated development based on the Blasius correlation (4.5), $f_{Blasius}$. The plot can be interpreted as comparison of the friction behavior of the non-circular duct with that of a cylindrical pipe having the same hydraulic diameter (4.6). For the presently considered hydraulic Reynolds number, no significant deviation in the friction factor was observed. The comparison can also be made for the remaining duct shapes and the results for

$$\Delta f = 1 - \frac{f}{f_{Blasius}} \tag{10.1}$$

are summarized in Table 10.1. Clearly, a reduction of the dimensionless friction factor can only be observed for geometry (d), which consists of very narrow corner regions. This result is in agreement with findings for triangular ducts, where only those having small apex angles lead to a deviation from the Blasius correlation (see Figure 9.1).

Table 10.1: Comparison of the friction factor and the pumping power resulting from turbulent flow in the non-circular ducts (a)–(d) at $Re_h = 4500$ with circular pipe flow. The duct shapes are referenced according to Figure 9.6.

case	Δf [%]	ΔP_P [%]
(a)	≈ 0	-31
(b)	≈ 0	-28
(c)	≈ 0	-15
(d)	15	-22

These considerations for the friction factor are based on the comparison of flows having the same hydraulic Reynolds number. It is fixed by keeping $U_b D_h$ constant. Consequently, the flow rates through these ducts are different. For the estimation of possible energy savings arising from corner regions, the pumping power that needs to be applied to transport a fluid through a non-circular duct will be compared with the pumping power that is necessary to transport this fluid with the same volume flow rate through a cylindrical pipe. For this purpose, the length scale D_m is introduced, which denotes the diameter of a circular-shaped pipe having the same cross section area, A_{cs}, as the modified duct:

$$D_m = 2\left(\frac{A_{cs}}{\pi}\right)^{1/2} \quad . \tag{10.2}$$

Using the factor D_m/D_h, which depends on the non-circular duct shape considered and is always positive, the previously introduced friction factors can be used together with the relationships given in Equations (4.3) and (4.10) to evaluate differences in the pumping power:

$$\Delta P_P = 1 - \frac{P_{P,\,\text{(a)–(d), turbulent}}}{P_{P,\,\text{circular, turbulent}}} = 1 - \frac{f}{f_{Blasius}}\left(\frac{D_m}{D_h}\right)^{5/4}$$
$$\text{with} \quad \frac{D_m}{D_h} = \frac{C}{2\,(A_{cs}\,\pi)^{1/2}} \quad , \tag{10.3}$$

where C is the perimeter of the non-circular duct. According to this relationship, the pumping power resulting from turbulent flow through each duct geometry (a)–(d) is significantly increased compared with the cylindrical pipe leading to negative values for ΔP_P (see Table 10.1).

It should be noted that for turbulent flow in case (b) a progressive trend in f to lie below the Blasius correlation for increasing Reynolds numbers was observed in Figure 9.14. Extrapolating this tendency, a positive result for ΔP_P and thus

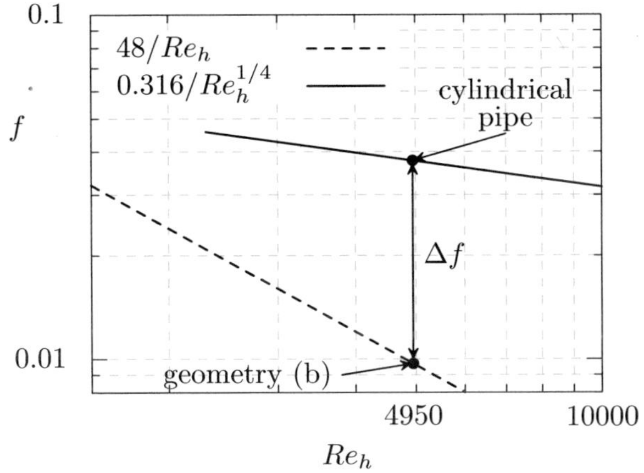

Figure 10.1: Numerical results for the friction factor f arising for geometry (b) and the cylindrical pipe at $Re_h = 4950$. The same initial conditions are used for the simulations [72]. The laminar friction law, $f = 48/Re_h$, for the flow through geometry (b) and the Blasius correlation [106] are included.

a reduction of the pumping power compared with the cylindrical pipe might be possible for very high Reynolds numbers, namely beyond $Re_h = 170000$.
The investigation of physical properties in the turbulent velocity field in Section 9.2 suggested that the proposed modified duct shapes have potential to ensure laminar flow under conditions where it can be expected to be turbulent in the cylindrical pipe. In the simulations discussed earlier, the flow is initially significantly disturbed. For less disturbed conditions, direct numerical simulations performed by Krieger [72] indicate that a pipe bounded by a wavy contour (see geometry (b) in Figure 9.6) leads to laminar flow for certain initial conditions, which are found to result in turbulent flow in a cylindrical pipe [72]. In this comparison the hydraulic Reynolds number is fixed to $Re_h = 4950$. The friction law for laminar flow in geometry (b) evaluates to $f = 48/Re_h$. Figure 10.1 shows that the friction factor of the laminar flow is significantly lowered compared with the turbulent case, namely by 75%. Similar to previous analysis, the set-up used in the simulations leads to different volume flow rates through the pipes with the various cross section shapes. Therefore, the reduction of the friction and, conclusively, the pressure drop at a fixed hydraulic Reynolds number due to geometrical modifications does not necessarily lead to benefits in the pumping power.

The evaluation of benefits in the pumping power due to the persistence of laminar flow in modified ducts is essential in order to evaluate the practical relevance of the proposed flow control strategy. Its calculation will be demonstrated next. The friction factor of a laminar, fully developed flow through a straight duct can generally be expressed by $f = a/Re_h$ (4.7), where the value of a depends on the cross section shape of the duct [106]. Following Equation (10.3), the pumping power can again be evaluated based on the relationship of $f/f_{Blasius}$:

$$\Delta P_P = 1 - \frac{P_{P,\,\text{(a)–(d), laminar}}}{P_{P,\,\text{circular, turbulent}}} = 1 - \frac{f}{f_{Blasius}} \left(\frac{D_m}{D_h}\right)^{5/4} = 1 - \frac{a}{0.316\, Re_h^{3/4}} \left(\frac{D_m}{D_h}\right)^{5/4} \quad \text{with} \quad \frac{D_m}{D_h} = \frac{C}{2\,(A_{cs}\,\pi)^{1/2}} \quad . \tag{10.4}$$

The use of D_m corresponds to the evaluation of f at a Reynolds number at which the volume flow rate through the cylindrical pipe is equal to that in the non-circular duct. The difference in the friction factor at this Reynolds number is proportional to ΔP_P and is illustrated in Figure 10.2 as Δf^*. Obviously, the benefit in the pumping power is significant if the flow stays laminar.

According to Equation (10.4), the possibility to reduce the pumping power due to ensuring laminar flow in modified ducts depends on the duct shape, represented by a characteristic value of a and D_m/D_h, and on the hydraulic Reynolds number. Using the proposed procedure, the benefit in the pumping power for different pipes can be quantified. The results at the Reynolds number considered by Krieger, $Re_h = 4950$, together with the individual laminar friction laws are summarized in Table 10.2. It is evident that the possible gain in the pumping power is large for all duct geometries considered.

Figure 10.2 illustrates that the difference in the friction factor and the pumping power due to the persistence of laminar flow in the modified duct geometry increases with increasing Reynolds number. From the dependence of f on Re_h, it can be seen that the friction factor f in the laminar regime is lower than in turbulence for the entire range of Reynolds numbers where the flow might reach the turbulent state, *i.e.*, $Re_h > 2300$. In contrast, for Δf^*, and consequently for ΔP_P, a similar universal statement cannot be directly extracted from the plot. The border for the hydraulic Reynolds number to achieve reduction of the pumping power compared with turbulent flow in a cylindrical pipe gives rise to

$$Re_h > \left(\frac{D_m}{D_h}\right)^{5/3} \left(\frac{a}{0.316}\right)^{4/3} \quad \text{with} \quad \frac{D_m}{D_h} = \frac{C}{2\,(A\pi)^{1/2}} \quad . \tag{10.5}$$

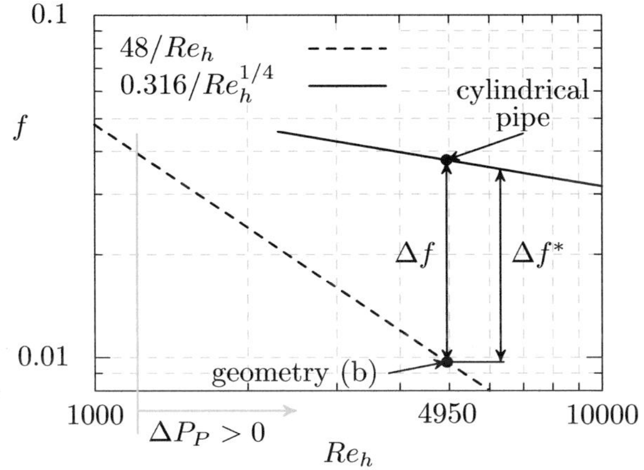

Figure 10.2: Reduction in the friction factor for laminar flow in geometry (b) at $Re_h = 4950$ compared with turbulent flow in the cylindrical pipe. For the evaluation of Δf^* the two pipes have the same cross section area. Δf^* is proportional to the differences in the pumping power. The hydraulic Reynolds number that has to be reached in order to reduce P_P is also indicated. Numerical results from Krieger [72], the laminar friction law for the flow through geometry (b) and the Blasius correlation [106] are included.

The previous equation is again derived for equal volume flow rates in the ducts. The limiting values for Re_h evaluated for the different ducts are summarized in Table 10.2. They are below the critical value after which transition to turbulence is expected, which is illustrated as an example for geometry (b) in Figure 10.2. These results indicate that the delay of the transition to turbulence due to similar modifications of the duct geometry, as considered here, will lead to benefits in the pumping power in the entire Reynolds number range in which the flow in the cylindrical pipe can be expected to be turbulent.

The high energy savings which result from the proposed flow control technique might be attractive for various application areas: the transport of fluid in pipe systems is ubiquitous in many industrial and civil areas. Notable examples are found in the chemical industry and in public water supplies. In the context of the aimed practical application of ducts with non-circular cross section shape it should be noted that the manufacturing of these ducts can be expected to be accompanied by increased costs. The costs, together with possible novel manufacturing techniques, have to be carefully considered.

Table 10.2: Friction laws for laminar flow in the non-circular ducts presented in Figure 9.6, possible reduction of the pumping power at $Re_h = 4950$ compared with turbulent flow in a cylindrical pipe and the Reynolds number that has to be reached in order to obtain a reduction in the pumping power.

case	$f_{laminar}$	ΔP_P [%]	$Re_{h\,limit}$
(a)	$48/Re_h$	63	1180
(b)	$48/Re_h$	65	1220
(c)	$54/Re_h$	64	1160
(d)	$40/Re_h$	67	1040

However, these costs only have to be accounted for once. The possible positive economics of the permanent operation of pipe systems are significant and can be expected to overcome the additional effort that is necessary for the installation.

11 Conclusion and outlook

An investigation into the resistance of laminar and turbulent flows was undertaken in the present work. In this context, stationary and fully developed flows through straight ducts are considered, the entire dissipation of which can be balanced by friction forces acting on the wall. Physical models are developed that theoretically can lead to reduction of the energy dissipated by the fluid motion and thus potentially provide energy savings. An attempt is made to establish the relevant mechanisms due to the appropriate design of internal flow domains.
In a laminar regime, the flow is solely governed by viscous effects and the cylindrical pipe can be proven to minimize the energy dissipation per volume flow rate [107]. The optimality conditions of the pure circular cross section shape are accompanied by the fact that they minimize the wetted perimeter compared with the cross section area. In contrast, if the height of the flow domain is restricted and channel flow is considered, unstructured walls are no longer optimal. Despite the increase in the wetted surface, wall structures that are wide compared with their height are shown to reduce the overall friction losses substantially. These results highlight the major impact of variations in the channel height on the flow resistance and suggest that using the available installation space in this respect has high potential for energy savings in practical applications.
If the Reynolds number increases and the flow turns to the turbulent state, the energy dissipated by the fluid motion increase significantly. Thus, ensuring the laminar flow regime under conditions at which the flow is typically found to be turbulent has a large potential for decreasing the flow resistance.
The application of statistical tools enables the derivation of related transport equations for transitional and turbulent flow [67, 103]. From this description of the flow it is found that provoking axisymmetry in the stresses leads to suppression of the dissipation originating from the fluctuating motion, and thus this is a common goal for the delay of transition to turbulence and flow laminarization [65, 66]. The potential of modifications to duct cross section shapes to provoke the theoretically suggested mechanism is investigated in direct numerical simulations.
In this context, coexisting laminar and turbulent regions appearing in narrow corner areas of ducts are studied. Examination of the statistical properties of the flow shows that the stability of the laminar flow region and the process of flow

laminarization obey the theoretically derived mechanism. Corner regions in ducts are shown to provoke the intended tendency towards axisymmetry.
Further, the coexistence of laminar and turbulent flows allows the investigation of the transition between the flow states. Based on the statistical description of the flow, an attempt is made to contribute to the general understanding of this process, which is of considerable scientific interest. The breakdown to turbulence is observed to be associated with a decrease in the anisotropy and axisymmetry in the stress tensor. This finding for the transition within the cross section plane of non-circular ducts could be discussed further and possibly generalized in connection with the transition process arising in, for example, a cylindrical pipe or flat plate boundary layer flow.
The impact of corner regions on the statistical properties of turbulent flow is investigated further. In order to do so, duct cross section shapes composed of multiple corner regions are developed. To extend the axisymmetrical state of the stresses to a large part of the flow, the scale of the corresponding wall structures is significantly increased compared with that proposed for example for riblets, which are known to mainly act in the near-wall region [21]. The design of the corner and crest regions is varied within the investigation. In general, a tendency towards axisymmetry is observed in the proximity of the corner bisector, which causes vanishing turbulent dissipation but a deviation from this state is present above the crest. However, experimental results in the literature [105] suggest that the beneficial effect arising from corners prevails and this type of duct geometries is recommended being considered for the present purposes. This investigation shows that acute corner angles and profiled connecting walls optimize the overall behavior. Future numerical work might further consider the preferred shaping of ducts in investigating the limiting angle in the corner that is necessary to provoke axisymmetry in the stresses. Since acute corners in internal flow domains are difficult to manufacture, this fact is of importance for the practical implementation of the described method for friction drag reduction.
In general, the theoretical analysis together with the findings for the coexisting laminar and turbulent flow fields indicate that the same properties in the stresses lead to laminarization and delay of transition, suggesting that the novel ducts are suitable for both objectives. However, on further investigation of the mean flow field in these ducts, it is found that the benefits due to the axisymmetrical state of the stresses in the corner regions might be partially compensated. The considered duct geometries induce inflection instabilities in the mean velocity profile in the circumferential direction. Comparison of the inflection instability in a laminar and turbulent velocity profile indicates that persistence of laminar flow is more likely to be achieved than laminarization of turbulent flow. This argument is in

agreement with previous observations on the influence of surface morphology in the form of surface embedded grooves on transitional and turbulent flow [40, 64]. Crest regions in the form of smooth curvatures are assumed to limit the amount of instability and should be further considered for the improvement of passive control methods.

Based on the findings for particular duct geometries, it is concluded that the cylindrical pipe might not be optimal if higher Reynolds numbers are considered: corner regions have the potential to ensure laminar flow where the breakdown to turbulence is expected in a pipe with a circular cross section. The related energy savings are quantified and found to be substantially high, supporting the modified duct shapes for numerous practical applications.

In order to explore the efficiency of ducts that include corner regions in practice, their stabilizing effect has to be further assessed. For this purpose, experimental investigations are proposed. Suitable pipes are already available commercially (*e.g.*, Mendener Präzisionsrohr GmbH [1]). In these experiments, the impact of different initial conditions on the possibility for laminar flow to persist can be tested. These conditions, together with the Reynolds number under which the experiments are run, will be initiated from the boundary conditions found in practice.

Besides investigating the physical mechanisms in the flow field that foster or limit the potential for ensuring laminar flow, the impact of the present duct shapes on the dissipation in the turbulent regime is analysed. At the relatively low Reynolds number that is considered, the beneficial effects arising from the corners do not overcome the increase in the wetted perimeter and the energy that has to be applied in order to transport the flow is increased compared with the cylindrical pipe. However, results for higher Reynolds numbers suggest that the modification of the flow initiated by the cross section shape leads to an overall decrease in the turbulent dissipation, which continues to prevail the entire dissipation [76, 105]. Thus, the present type of configurations might have potential for energy savings in the high Reynolds number range that is important in practice, and which is reached, for example, in water supply pipelines, even if the flow cannot be kept in the laminar state.

For future work, the continuation of investigations on the impact of the duct geometry on turbulent flow is proposed. Experimental investigations should tackle the verification and possible extension of the discussed trend in the friction losses for high Reynolds numbers, which can not really be reached numerically. Additionally, the direct estimation of the turbulent dissipation within the cross section shape is of interest. In particular, its comparison for different Reynolds numbers and various cross section designs might support the physical understanding of turbulent drag reduction and lead to further optimization of cross section shapes in this respect. However, the calculation of the turbulent dissipation is expected to be a challenging issue for the numerical procedure applied within the present work. It can be assumed that the non-orthogonal nature of a polyhedral mesh influences the evaluation of the required gradients in the fluctuating field in a critical manner. Certainly, a carefully extended validation study is necessary, including the computation and balancing of all terms in the transport equation for the Reynolds stresses.

Bibliography

[1] MPG Mendener Praezisionsrohr GmbH, August 2012.

[2] http://www.akresource.org/curriculum_cd/energy/resources/alyeska.pdf, March 2014.

[3] http://www.flowchem-dra.com/how-drag-reduction-additives-work.html, March 2014.

[4] http://www.oilfieldwiki.com/wiki/drag_reducing_agent, March 2014.

[5] J. Alamo and J. Jiménez. Spectra of the very large anisotropic scales in turbulent channels. *Phys. Fluids*, 15(6):41–44, 2003.

[6] A. Aly, A. Trupp, and A. Gerrard. Measurements and prediction of fully developed turbulent flow in an equilateral triangular duct. *J. Fluid Mech.*, 85:57–83, 1978.

[7] R. Antonia, J. Kim, and L. Browne. Some characteristics of small-scale turbulence in a turbulent duct flow. *J. Fluid Mech.*, 233:369–388, 1991.

[8] M. Avila, A. Willis, and B. Hof. On the transient nature of localized pipe flow turbulence. *J. Fluid Mech.*, 646:127–136, 2010.

[9] M. Bahrami, M. Yovanovich, and J. Culham. A novel solution for pressure drop in singly connected microchannels of arbitrary cross-section. *Int. J. Heat Mass Transfer*, 50:2492–2502, 2007.

[10] P. Bandopadhayay and J. Hinwood. On the coexistence of laminar and turbulent flow in a narrow triangular duct. *J. Fluid Mech.*, 59:775–783, 1973.

[11] D. Bechert, M. Bruse, W. Hage, J. van der Hoeven, and G. Hoppe. Experiments on drag-reducing surfaces and their optimization with an adjustable geometry. *J. Fluid Mech.*, 338:59–87, 1997.

[12] T. Bewley. A fundamental limit on the balance of power in a transpiration-controlled channel flow. *J. Fluid Mech.*, 632:443–446, 2009.

[13] R. Blackwelder. Some ideas on the control of near-wall eddies. In *AIAA 2nd shear flow conference, Tempe, USA*, 1989.

[14] R. Blackwelder and R. Kaplan. On the wall structure of the turbulent boundary layer. *J. Fluid Mech.*, 76:89–112, 1976.

[15] M. Breuer. *Direkte Numerische Simulation und Large-Eddy Simulation turbulenter Strömungen auf Hochleistungsrechnern.* Habilitationsschrift, Friedrich-Alexander-Universität Erlangen-Nürnberg, Shaker Verlag, Aachen, 2002.

[16] E. Brundett and W. Baines. The production and diffusion of vorticity in duct flow. *J. Fluid Mech.*, 19:375–392, 1964.

[17] D. Carlson, S. Widnall, and M. Peeters. A flow-visualization study of transition in plane Poiseuille flow. *J. Fluid Mech.*, 121:487–505, 1982.

[18] L. Carlson and T. Irvine. Fully developed pressure drop in triangular shaped ducts. *J. Heat Transfer*, 83:441–444, 1961.

[19] C. Chin, A. Ooi, I. Marusic, and H. Blackburn. The influence of pipe length on turbulence statistics computed from direct numerical simulation data. *Phys. Fluids*, 22:115–107, 2010.

[20] H. Choi, P. Moin, and J. Kim. On the effect of riblets in fully developed laminar channel flows. *Phys. Fluids*, 3(8):1892–1896, 1991.

[21] H. Choi, P. Moin, and J. Kim. Direct numerical simulation of turbulent flow over riblets. *J. Fluid Mech.*, 225:503–539, 1993.

[22] G. Daschiel, T. Baier, J. Saal, and B. Frohnapfel. On the flow resistance of wide surface structures. *Proc. Appl. Math. Mech. (PAMM)*, pages 569–570, 2012.

[23] G. Daschiel, B. Frohnapfel, and J. Jovanović. Interpretation of the mechanism responsible for the persistance of a laminar region in turbulent duct flow. In *European turbulence conference (ETC14), Lyon, France*, 2013.

[24] G. Daschiel, B. Frohnapfel, and J. Jovanović. Numerical investigation of flow through a triangular duct: The coexistence of laminar and turbulent flow. *Int. J. Heat Fluid Flow*, 41(0):27–33, 2013.

[25] G. Daschiel, V. Krieger, J. Jovanović, and B. Frohnapfel. Designing smart duct geometries for low frictional losses. In *Proc. 66th Annual Meeting of the APS Division Fluid Dynamics, Pittsburgh, USA*, volume 58, 2013.

[26] G. Daschiel, M. Perić, J. Jovanović, and A. Delgado. The holy grail of microfluidics: sub-laminar drag by layout of periodically embedded microgrooves. *Microfluidics Nanofluidics*, 2013.

[27] A. Demuren and W. Rodi. Calculation of turbulence-driven secondary motion in non-circular ducts. *J. Fluid Mech.*, 140:189–222, 1984.

[28] A. Draad and F. Nieuwstadt. The earth's rotation and laminar pipe flow. *J. Fluid Mech.*, 577:297–308, 1998.

[29] F. Durst. *Grundlagen der Strömungsmechanik.* Springer-Verlag, Berlin, 2006.

[30] E. Eckert and T. Irvine. Flow in corners of passages with noncircular cross sections. *Trans. ASME*, 78(4):709–718, 1956.

[31] E. Eckert and T. Irvine. Pressure drop and heat transfer in a duct with triangular cross section. *J. Heat Transfer*, pages 709–718, 1960.

[32] J. Eggels, F. Unger, M. Weiss, J. Westerweel, R. Adrian, R. Friedrich, and F. Nieuwstadt. Fully developed turbulent pipe flow: a comparison between direct numerical simulation and experiment. *J. Fluid Mech.*, 268:175–209, 1994.

[33] O. Ertunć. personal communication. 2014.

[34] H. Faisst and B. Eckhardt. Sensitive dependence on initial conditions in transition to turbulence in pipe flow. *J. Fluid Mech.*, 504:343–352, 2004.

[35] F. Felten and T. Lund. Kinetic energy conservation issues associated with the collocated mesh scheme for incompressible flow. *J. Comp. Phys.*, 215:465–484, 2006.

[36] J. Ferziger and M. Perić. *Numerische Strömungsmechanik.* Springer-Verlag, Berlin, 2008.

[37] V. Fink. *Numerische Untersuchung der Beeinflussung der turbulent Kanalströmung durch Riblets.* Master thesis, Institut für Strömungsmechanik, Karlsruher Institut für Technologie, Karlsruhe, supervised by the author, 2014.

[38] B. Frohnapfel. *Flow control of near-wall turbulence.* Ph.D. thesis, Lehrstuhl für Strömungsmechanik, Friedrich-Alexander Universität Erlangen-Nürnberg, Erlangen, 2006.

[39] B. Frohnapfel, Y. Hasegawa, and M. Quadrio. Money versus time: evaluation of flow control in terms of energy consumption and convenience. *J. Fluid Mech.*, 700:406–418, 2012.

[40] B. Frohnapfel, J. Jovanović, and A. Delgado. Experimental investigation of turbulent drag reduction by surface-embedded grooves. *J. Fluid Mech.*, 590:107–116, 2007.

[41] B. Frohnapfel, P. Lammers, J. Jovanović, and F. Durst. Interpretation of the mechanism associated with turbulent drag reduction in terms of anisotropy invariants. *J. Fluid Mech.*, 577:457–466, 2007.

[42] K. Fukagata and N. Kasagi. Highly energy-conservative finite difference method for the cylindrical coordinate system. *J. Comput. Phys.*, 181:478–498, 2002.

[43] N. Fukushima and N. Kasagi. Turbulent momentum and heat transfer in ducts of rhombic cross section. *Proceedings of Heat transfer 2, Grenoble, France*, pages 207–212, 2002.

[44] S. Gavrilakis. Numerical simulation of low-Reynolds-number turbulent flow through a straight square duct. *J. Fluid Mech.*, 244:101–129, 1992.

[45] W. George and H. Hussein. Locally axisymmetric turbulence. *J. Fluid Mech.*, 233:1–23, 1991.

[46] F. Gessner and J. Jones. On some aspects of fully-developed turbulent flow in rectangular channels. *J. Fluid Mech.*, 23:689–713, 1965.

[47] N. Gilbert and L. Kleiser. Turbulence model testing with the aid of direct numerical simulation. In *Proc. of 8th Symposium on turbulent shear flows, Munich, Germany*, 1991.

[48] G. Grek, V. Kozlov, and S. Titarenko. An experimental study of the influence of riblets on transition. *J. Fluid Mech.*, 315:31–49, 1996.

[49] G. Grötzbach. *Direkte numerische Simulation turbulenter Geschwindigkeits-, Druck- und Temperaturfelder*. Dissertation, Universität Karlsruhe (TH), KFK-Bericht 2426, Gesellschaft für Kernforschung M.B.H, Karlsruhe, 1977.

[50] A. Güttler. personal communication. 2014.

[51] K. Hanjalić and B. Launder. A Reynolds stress model of turbulence and its application to thin shear flows. *J. Fluid Mech.*, 52:609–638, 1972.

[52] J. Hartnett, J. Koh, and S. McComas. A comparison of predicted and measured friction factors for turbulent flow through rectangular ducts. *J. Heat Transfer*, 84:82–88, 1962.

[53] S. He and M. Seddighi. Turbulence in transient channel flow. *J. Fluid Mech.*, 715:60–102, 2012.

[54] J. Hinze. *Turbulence*. McGraw-Hill Book Company, New York, 1959.

[55] B. Hof, A. de Lozar, M. Avila, X. Tu, and T. Schneider. Eliminating turbulence in spatially intermittent flows. *Science*, 327(5972):1491–1494, 2010.

[56] P. Holmes, J. Lumley, and G. Berkooz. *Turbulence, coherent structures, dynamical systems and symmetry*. Cambridge university press, 1998.

[57] K. Horiuti. Fully developed 2-d channel flow, Thtlab DNS database, 1992.

[58] A. Huser and S. Biringen. Direct numerical simulation of turbulent flow in a square duct. *J. Fluid Mech.*, 257:65–95, 1993.

[59] K. Iwamoto, Y. Suzuki, and N. Kasagi. Reynolds number effect on wall turbulence: towards effective feedback control. *Int. J. Heat Fluid Flow*, 23:678–689, 2002.

[60] N. Jarrin, S. Benamadouche, D. Laurence, and R. Prosser. Synthetic-eddy-method for generating inflow conditions for large eddy simulations. *Int. J. Heat Fluid Flow*, 27:585, 2006.

[61] H. Jasak. *Error analysis and estimation for the finite volume method with applications to fluid flows*. Ph.D. thesis, Department of mechanical engineering, Imperial college of science, technology and medicine, London, 1996.

[62] O. Jones. An improvement in the calculation of turbulent friction in rectangular ducts. *J. Fluids Eng.*, pages 173–180, 1976.

[63] J. Jovanović. *The statistical dynamics of turbulence*. Springer Verlag, Berlin, 2004.

[64] J. Jovanović, B. Frohnapfel, H. Lienhart, and A. Delgado. Experimental investigation of viscous drag reduction by flow control of laminar to turbulent transition using microgroove surface pattern. In *ETMM-8 Conference, Marseille, France*, pages 763–768, 2010.

[65] J. Jovanović, B. Frohnapfel, E. Škaljić, and M. Jovanović. Persistence of the laminar regime in a flat plate boundary layer at very high Reynolds number. *Thermal Sci.*, 10(2):63–96, 2006.

[66] J. Jovanović and R. Hillerbrand. On peculiar property of the velocity fluctuations in wall-bounded flows. *Thermal Sci.*, 9(1):3–12, 2005.

[67] J. Jovanović and M. Pashtrapanska. On the criterion for the transition onset and breakdown to turbulence in wall-bounded flows. *J. Fluids Eng.*, 126:626–633, 2004.

[68] J. Jovanović, M. Pashtrapanska, B. Frohnapfel, F. Durst, J. Koskinen, and K. Koskinen. On the mechanism responsible for turbulent drag reduction by dilute addition of high polymers: theory, experiments, simulations and predictions. *J. Fluids Eng.*, 128:118–130, 2006.

[69] J. Kim, P. Moin, and R. Moser. Turbulence statistics in fully developed channel flow at low Reynolds number. *J. Fluid Mech.*, 177:133–166, 1987.

[70] S. Kline, W. Reynolds, F. Schraub, and P. Runstadler. The structure of turbulent boundary layers. *J. Fluid Mech.*, 30:741–773, 1967.

[71] M. Kramer. Einrichtung zur Verminderung des Reibungswiderstandes. *Reichspatentamt, Patentschrift Nr. 669897*, 1937.

[72] V. Krieger. *Direkte numerische Simulation der Strömung durch ein Rohr mit sternförmigen Querschnitt - Eine Wiederaufnahme von Schillers zukunftsweisendem Experiment in Hinblick auf die Reibungsreduktion.* Bachelor thesis, Lehrstuhl für Strömungsmechanik, Friedrich-Alexander Universität Erlangen-Nürnberg, Erlangen, co-supervised by the author, 2012.

[73] A. Kuroda. *Direct Numerical Simulation of Couette-Poiseuille Flows.* Ph.D. thesis, The University of Tokyo, Tokyo, 1990.

[74] A. Kuroda and N. Kasagi. Fully developed 2-d channel flow, Thtlab DNS database, 1989.

[75] A. Kuroda, N. Kasagi, and M. Hirata. A direct numerical simulation of the fully developed turbulent channel flow. *International Symposium on Computational Fluid Dynamics, Nagoya*, pages 1174–1179, 1989.

[76] F. Laadhari. Reynolds number effect on the dissipation function in wall-bounded flows. *Phys. Fluids*, 19(3), 2007.

[77] D. Ladd, J. Rohr, L. Reidy, and E. Hendricks. The effect of riblets on laminar to turbulent transition. *Exp. Fluids*, 14:1–9, 1993.

[78] P. Lammers, J. Jovanović, B. Frohnapfel, and A. Delgado. Erlangen pipe flow: the concept and DNS results for microflow control of near-wall turbulence. *Microfluidics Nanofluidics*, 2012.

[79] G. Lemoult, J.-L. Aider, and E. Wesfreid. Turbulent spots in a channel: large-scale flow and self-sustainability. *J. Fluid Mech.*, 731, 2013.

[80] H. Lienhart, M. Breuer, and C. Köksoy. Drag reduction by dimples? - a complementary experimental/numerical investigation. *Int. J. Heat Fluid Flow*, 29:783–791, 2008.

[81] K. Liu, C. Christodoulou, O. Riccius, and D. Joseph. Drag reduction in pipes lined with riblets. *AIAA Journal*, 28, 1990.

[82] J. Lumley and G. Newman. The return of isotropy of homogeneous turbulence. *J. Fluid Mech.*, 82:161–178, 1977.

[83] T. Min, S. Kang, J. Speyer, and J. Kim. Sustained sub-laminar drag in a fully developed channel flow. *J. Fluid Mech.*, 558:309–318, 2006.

[84] R. Mises. *Elemente der Technische Hydrodynamik, 1. Teil.* , Leipzig, 1914.

[85] A. Mohammadi and J. Floryan. Pressure losses in grooved channels. *J. Fluid Mech.*, 725:23–54, 2013.

[86] P. Moin and K. Mahesh. Direct numerical simulation: a tool in turbulence research. *Ann. Rev. of Fluid Mech.*, 30(1):539–578, 1998.

[87] A. Monin and A. Yaglom. *The statistical fluid mechanics*, volume 1. MIT Press, Cambridge, 1987.

[88] R. Moser, J. Kim, and N. Mansour. Direct numerical simulation of turbulent channel flow up to $Re_\tau = 590$. *Phys. Fluids*, 11(4):943–945, 1999.

[89] S. Nakao. Application of V shape riblets to pipe flow. *J. Fluids Eng.*, 113:587–590, 1991.

[90] A. Nakayama, W. Chow, and D. Sharma. Calculation of fully developed turbulent flows in ducts of arbitrary cross-section. *J. Fluid Mech.*, 128:199–217, 1983.

[91] N. Nikitin and A. Yakhot. Direct numerical simulation of turbulent flow in elliptical ducts. *J. Fluid Mech.*, 532:141–164, 2005.

[92] J. Nikuradse. Turbulente Strömungen in nicht-kreisförmigen Rohren. *Ing.-Arch.*, 1:306–332, 1930.

[93] M. Nishi. *Laminar to turbulent transition in pipe flow through puffs and slugs.* Ph.D. thesis, Lehrstuhl für Strömungsmechanik, Friedrich-Alexander Universität Erlangen-Nürnberg, Erlangen, 2009.

[94] R. Perić. *Direct numerical simulation of turbulent channel flow with longitudinal riblets on one wall - Assessment of drag reduction.* Bachelor thesis, Lehrstuhl für Strömungsmechanik, Friedrich-Alexander Universität Erlangen-Nürnberg, Erlangen, 2011.

[95] H. Perkins. The formation of streamwise vorticity in turbulent flow. *J. Fluid Mech.*, 44:721–740, 1970.

[96] W. Pfenniger. Transition in the inlet length of turbes at high Reynolds numbers. *Boundary layer and flow control*, pages 970–980, 1961.

[97] A. Pinelli, M. Uhlmann, A. Sekimoto, and G. Kawahara. Reynolds number dependence of mean flow structure in square duct turbulence. *J. Fluid Mech.*, 644:107–122, 1 2010.

[98] O. Pironneau and G. Arumugam. On riblets in laminar flow. *Control of boundaries and stabilization*, pages 51–65, 1989.

[99] S. Pope. *Turbulent flows.* Cambridge University Press, Cambridge, 2011.

[100] T. Pöschl. Zweiter Internationaler Kongreß für Technische Mechanik in Zürich (12. bis 17. September 1926). *Naturwissenschaften*, 14:1029–1032, 1926.

[101] R. Radeztsky and S. R. an W. Saric. Effect of isolated micron-sized roughness on transition swept-wing flows. *AIAA Journal*, 37, 1999.

[102] H. Raiesi, A. Pollard, and U. Piomelli. Direct numerical simulations of turbulence induced secondary motion in square and skewed ducts. *Proceedings of TSFP-7, Ottawa, Canada*, 2011.

[103] O. Reynolds. On the dynamical theory of incompressible viscous fluids and the determination of the criterion. *Philosophical Transactions of the Royal Society of London A*, 186:123–164, 1894.

[104] J. Rotta. *Turbulente Strömungen*. B. G. Teubner, Stuttgart, 1972.

[105] L. Schiller. Über den Strömungswiderstand von Rohren verschiedenen Querschnitts und Rauhigkeitsgrades. *Zeitschr. Appl. Math. Mech. (ZAMM)*, 3:2–13, 1923.

[106] H. Schlichting. *Boundary-layer theory*. McGraw-Hill Book Company, New York, 7th edition, 1979.

[107] A. Schulz. The optimal shape of a pipe. *Zeitschr. Appl. Math. Mech. (ZAMM)*, 64:1177–1185, 2013.

[108] P. Spalart. Numerical study of sink-flow boundary layers. *J. Fluid Mech.*, 172:307–328, 1986.

[109] P. Spalart. Direct simulation of a turbulent boundary layer up to $R_\theta = 1410$. *J. Fluid Mech.*, 187:61–98, 1988.

[110] H. Tennekes and J. Lumley. *A first course in turbulence*. MIT Press, Cambridge, 1972.

[111] M. Uhlmann, A. Pinelli, G. Kawahara, and A. Sekimoto. Marginally turbulent flow in a square duct. *J. Fluid Mech.*, 588:153–162, 2007.

[112] F. Unger and R. Friedrich. Fully developed turbulent pipe flow, Thtlab DNS database, 1993.

[113] E. Villiers. *The potential of large eddy simulation for the modelling of wall bounded flows*. Ph.D. thesis, Department of mechanical engineering, Imperial college of science, technology and medicine, London, 2006.

[114] F. White. *Viscous fluid flow*. McGraw-Hill Book Company, New York, 1974.

Nomenclature

All indices used (i, j, k,...) run from 1-3. The superscript "+" indicates that the quantity is normalized with the wall shear velocity u_τ and the kinematic viscosity ν.

Latin letters

upper case

symbol	SI unit	description
II, III	-	scalar invariants of the anisotropy tensor a_{ij}
A	-	measure for axisymmetry
B	-	constant, $B \simeq 5.2$
$B(\varphi)$	m	length of circular arc
A_{cs}	m^2	cross section area
A_w	m^2	lateral surface
C	m	perimeter
Co	-	Courant number
D_m	m	diameter of circular pipe
D_h	m	hydraulic diameter
Eu	-	Euler number
F, G	-	scalar functions
H	m	duct height
$2L$	m	mean channel height
L_{x_1}	m	streamwise extension of the low domain
P	$kg/(s^2\,m)$	instantaneous pressure
P'	$kg/(s^2\,m)$	time averaged pressure of laminar base flow
$\overline{P}$	$kg/(s^2\,m)$	time averaged pressure of turbulent flow field
ΔP	$kg/(s^2\,m)$	pressure difference between outlet and inlet

P_k	$m^2/(s^3)$	production of kinetic energy of turbulence
P_P	$kg/(s^2\,m)$	pumping power
R_b	m	length of corner bisector
R_λ	-	turbulent Reynolds number
Re	-	Reynolds number
Re_h	-	hydraulic Reynolds number
Re_τ	-	friction Reynolds number
St	-	Strouhal number
U_i	m/s	instantaneous flow velocity
U_i'	m/s	time averaged velocity of laminar base flow
$\overline{U}_i$	m/s	time averaged velocity of turbulent flow field
U_b	m/s	bulk flow velocity
U_z	m/s	laminar streamwise velocity
$\overline{U}_z$	m/s	time averaged turbulent streamwise velocity
V	m^3	volume of the flow domain
$\dot{V}$	m^3/s	flow rate

lower case

symbol	SI unit	description
a, b	m	variables describing structure geometry
a_1, a_2	-	coefficients of Taylor series expansion
a_{ij}	-	anisotropy tensor
b_1, b_2	-	coefficients of Taylor series expansion
c_1, c_2	-	coefficients of Taylor series expansion
f	-	friction factor
f	s^{-1}	frequency
f_i	m/s^2	acceleration
f_κ	s^{-1}	Kolmogorov frequency scale
$k=q^2/2$	m^2/s^2	kinetic energy of turbulence
$2l$	m	channel width
p	$kg/(s^2\,m)$	instantaneous turbulent pressure fluctuation

p'	kg/(s^2 m)	instantaneous pressure disturbance
r_b	m	length of wall/crest bisector
s	m	spacing of riblets
t	s	time
u_τ	m/s	wall shear velocity
u_i	m/s	instantaneous turbulent velocity fluctuation
$u_{i,\mathrm{rms}}$	m/s	root mean square of the velocity fluctuations
$u_{i,\mathrm{rms}}$	m/s	root mean square of the velocity disturbances
u'_i	m/s	instantaneous velocity disturbance
$\overline{u_i u_j}$	m^2/s^2	Reynolds stresses
$\overline{u'_i u'_j}$	m^2/s^2	apparent stresses
u_K	m/s	Kolmogorov velocity scale
x_i	m	Cartesian coordinates
z, r, φ	m	Cylindrical coordinates

Greek letters

upper case

symbol	SI unit	description
Φ	m^2/s^3	dissipation per unit mass and time
$\langle\Phi\rangle$	$kg\,m^2/s^3$	entire dissipation rate of the working fluid

lower case

symbol	SI unit	description
α	$^\circ$	corner angle
δ_{ij}	-	Kronecker delta
δ	m	channel half height, pipe radius
ϵ	m^2/s^3	turbulent dissipation
ϵ_d	m^2/s^3	direct dissipation
η_K	m	Kolmogorov length scale
κ	-	constant, $\kappa \simeq 0.41$
κ	m^{-1}	wave number
λ_i	-	unit vector
μ	kg/(m s)	dynamic viscosity of the fluid
ν	m^2/s	kinematic viscosity of the fluid
ρ	kg/m^3	density of the fluid
τ_{ik}	$kg/(m\,s^2)$	stress tensor
τ_w	$kg/(m\,s^2)$	wall shear stress

Abbreviations

symbol	description
DNS	Direct Numerical Simulation
DR	Drag Reduction
DRA	Drag Reducing Additive
hex	hexahedral
PISO	Pressure Induced with Splitting of Operator
poly	polyhedral

List of Figures

4.1 Straight duct with arbitrary cross section shape that is homogeneous in the streamwise x_1-direction. L_{x_1} denotes the length of the considered flow domain. 10
4.2 Dependence of the friction factor on the Reynolds number for laminar and turbulent flow through a smooth cylindrical pipe. 12

5.1 Numerically predicted laminar flow through a flat channel and channel with wide surface structures. 20
5.2 Contour plot of the drag reduction that can be achieved with structured walls compared with a flat reference channel in laminar flow. 22

6.1 Anisotropy–invariant map [82]. 32
6.2 Turbulent dissipation rate at the wall *versus* the magnitude of anisotropy at the wall following the illustration of Jovanović and Hillerbrand [67]. 35
6.3 Physical model aiming at low friction losses and persistence of the laminar regime in wall-bounded flows. 37
6.4 Physical model aiming at the reduction of energy dissipation and laminarization of turbulent flow. 41
6.5 Sketch of the riblet geometry and anisotropy invariant mapping of the stresses in turbulent channel flow over riblets. 44
6.6 Anisotropy invariant mapping for the stresses arising from DNS of turbulent pipe and channel flow for the same friction Reynolds number $Re_\tau = 180$. 46
6.7 Procedure for the investigation of low energy dissipation in duct flows based on theoretical analysis for laminar and turbulent flow. 48

7.1 The lower left quadrant of the computational grids consisting of hexahedral (left) and polyhedral (right) cells used for the DNS of the turbulent square duct flow. 55
7.2 The computational domain for the simulation of turbulent flow in a square duct. 55

7.3 Development of the rms-values of the normal Reynolds stresses normalized with U_b at two different spanwise positions: $x_3/H = 0.15$ (upper figure), $x_3/H = 0.35$ (lower figure). 58
7.4 Illustration of the mechanism and the magnitude of the secondary motion together with the resulting distribution of the friction on the wall for the square duct. 60
7.5 Development of the secondary flow component $\overline{U}_3$ at the spanwise position $x_3/H = 0.15$. 61
7.6 Development of the streamwise mean flow component $\overline{U}_1$ at the spanwise positions $x_3/H = 0.15$ (upper figure) and $x_3/H = 0.5$ (lower figure). 62
7.7 Development of the wall shear stress along the duct walls. 63
7.8 The friction factor resulting from square duct flow: comparison of analytical and empirical relationships with data from simulations with different grids and duct lengths. 64
7.9 Sketch of the isosceles triangular duct with an apex angle of 11.5° where x_1 is the main flow direction (left) and the mesh in the corner region of the duct, namely up to $x_3/H \approx 0.1$. 66
7.10 Frequency spectrum normalized with Kolmogorov scales arising at the position $(x_2, x_3) = (0.0, 0.7H)$. 67
7.11 One-dimensional energy spectrum of the streamwise wavenumber normalized with Kolmogorov scales arising at the position $(x_2, x_3) = (0.0, 0.7H)$. 68
7.12 Development of the mean flow component $\overline{U}_1$ along the symmetry line of the triangular duct. 69
7.13 Development of the mean flow component $\overline{U}_1$ at the spanwise positions $x_3/H = 0.25$, $x_3/H = 0.3$ and $x_3/H = 0.5$ together with sketches of parabolic profiles. 69

8.1 Cross section shapes provoking coexisting laminar and turbulent flow regions. 72
8.2 Contour plot of the mean streamwise velocity field $\overline{U}_1$ normalized with U_b. 74
8.3 Contour plot of the magnitude of the secondary flow field. 75
8.4 Mean streamwise velocity, $\overline{U}_1$, normalized with the local wall shear velocity, u_τ, plotted along the x_2-direction at different spanwise positions. 76
8.5 Wave number spectrum normalized with Kolmogorov scales at different spanwise positions on the x_3-axis of the triangular duct (a). 77
8.6 Dissipation spectrum normalized with Kolmogorov scales at different spanwise positions on the x_3-axis of the triangular duct (a). 78

8.7 Reynolds stresses $\overline{u_i u_j}$ normalized with the local wall shear velocity u_τ plotted along the x_2-direction at different spanwise positions together with the profiles for plane channel flow at similar values of Re_τ. 80
8.8 Apparent stresses $\overline{u_i' u_j'}$ normalized with the local wall shear velocity u_τ plotted along the x_2-direction at different spanwise positions. . . . 81
8.9 Trajectory through the anisotropy invariant map along the x_2-direction at different spanwise positions in the turbulent region. 82
8.10 Trajectory through the anisotropy invariant map along the x_2-direction at different spanwise positions in the laminar region. 83
8.11 Development of the normalized rms-values of the velocity fluctuations $u_{i,rms}$ along the x_3-axis of the triangular duct with 11.5° apex angle. . . 85
8.12 Trajectory through the anisotropy invariant map along the x_3-axis of the duct. 86
8.13 Development of the magnitude of anisotropy, $\mathrm{II}/\mathrm{II}_{\mathrm{max}}$, and of the measure for axisymmetry, A, along the x_3-axis of the triangular duct with 11.5° apex angle. 87
8.14 Development of the magnitude of anisotropy and of the measure for axisymmetry along the x_3-axis of the triangular duct with 11.5° apex angle and in circular pipe flow at $Re_h = 4950$. 90
8.15 Development of the mean streamwise velocity, $\overline{U}_1$ along the x_3-axis of the triangular duct with 11.5° apex angle. 91
8.16 Development of the normalized rms-values of the velocity fluctuations $u_{i,\mathrm{rms}}$ along the x_3-axis of the triangular duct with the 4° apex angle. . 93
8.17 Development of the magnitude of anisotropy, $\mathrm{II}/\mathrm{II}_{\mathrm{max}}$, and of the measure for axisymmetry, A, along the x_3-axis of the triangular duct with the 4° apex angle. 94
8.18 Development of the normalized rms-values of the velocity fluctuations $u_{i,\mathrm{rms}}$ along the corner bisector of the diamond shaped duct. 96
8.19 Development of the magnitude of anisotropy, $\mathrm{II}/\mathrm{II}_{\mathrm{max}}$, and of the measure for axisymmetry, A, along the x_3-axis of the diamond shaped duct. 97
8.20 Contour plot of the magnitude of the secondary flow field normalized with U_b together with the secondary flow $\overline{U}_3$ developing in a corner of the diamond shaped duct. 99
8.21 Contour plot of the magnitude of the secondary flow field normalized with U_b in a section of the triangular ducts with 4° and 11.5° apex angle. 99

9.1 Friction factor f plotted against the hydraulic Reynolds number Re_h for different non-circular duct geometries. 103

9.2 Cross section plane of the square duct and equilateral triangular duct with corner and wall bisectors. 106
9.3 Reynolds stresses along the wall bisector and along the corner bisector of the square duct. 108
9.4 Reynolds stresses along the wall bisector and along the corner bisector of the equilateral triangular duct. 109
9.5 Trajectory through the anisotropy invariant map along the wall and corner bisectors of the square and the equilateral triangular duct. . . . 110
9.6 Four novel duct geometries with differently designed tip and crest regions. 113
9.7 Cross section plane of the star pipe with crest and corner bisector. . . 114
9.8 Reynolds stresses along the crest bisector and along the tip bisector of the pipe with star-shaped cross section. 115
9.9 Reynolds stresses along the crest bisector and along the tip bisector of case (b). 117
9.10 Reynolds stresses along the crest bisector and along the tip bisector of case (c). 118
9.11 Reynolds stresses along the crest bisector and along the tip bisector of case (d). 119
9.12 Trajectory through the anisotropy invariant map along the crest and tip bisectors of cases (a) and (b). 121
9.13 Trajectory through the anisotropy invariant map along the crest and tip bisectors of cases (c) and (d). 122
9.14 Friction factor f plotted over the hydraulic Reynolds number Re_h for geometry (b) [105] 124
9.15 Production of kinetic energy of turbulence, P_k, in the cross section of duct (b) and a cylindrical pipe. 125
9.16 Production of kinetic energy of turbulence, P_k, in the duct cross sections. 126
9.17 Extension and application of Jovanović and Hillerbrand's [67] illustration of the turbulent dissipation rate at the wall *versus* the magnitude of anisotropy at the wall for analyses of non-circular duct flows. . . . 127
9.18 Segment of the cross section plane of geometry (b) with a cylindrical coordinate system. 130
9.19 Streamwise mean velocity profiles along circular arcs and along corner bisectors for turbulent flow at $Re_h = 4500$ in geometry (b). 131
9.20 Shear stresses $\overline{u_z u_r}$ and $\overline{u_z u_\varphi}$, in the cross section of duct (b). 132
9.21 Shear of the streamwise velocity profiles $\overline{U}_z$ in the circumferential direction along circular arcs in geometry (b). 133

9.22 Streamwise laminar and turbulent velocity profiles along circular arcs in geometry (b). 134
9.23 Shear in the circumferential direction of the laminar and turbulent streamwise velocity profiles in geometry (b). 135

10.1 Numerical results for the friction factor f arising for geometry (b) and the cylindrical pipe at $Re_h = 4950$. 139
10.2 Evaluation of savings in the pumping power for laminar flow in geometry (b) compared with turbulent flow in the cylindrical pipe. 141

List of Tables

7.1 Standard domain size and spatial discretization where the viscous units are calculated using the fluid viscosity and the friction velocity estimated *a priori* with the Blasius correlation [106]. 52
7.2 DNS of turbulent flow in a square duct: domain size and spatial discretization in the current study and in the reference cases. 56
7.3 Spatial discretization close to the wall at the most critical positions in the current study's square duct cases. 56
7.4 DNS of turbulent flow in the triangular duct with 11.5° apex angle: domain size and spatial discretization. 66
7.5 Spatial discretization in the triangular duct with an apex angle of 11.5° based on Kolmogorov length scale. 66

8.1 Spatial discretization close to the wall at the most critical position in the flow domains with coexisting laminar and turbulent flow fields. . . 73
8.2 Local Reynolds number arising at the origin point (ii) and at the transition point (iii) within the different duct geometries. 98

10.1 Comparison of the friction factor and the pumping power resulting from turbulent flow in different non-circular ducts at $Re_h = 4500$ to circular pipe flow. 138
10.2 Friction laws for laminar flow in different non-circular ducts, possible reduction of the pumping power at $Re_h = 4950$ and Reynolds-number, which has to be reached in order to obtain a reduction in the pumping power. 142

A.1 Drag reduction in channel flow due to wide surface structures: Comparison of model, numerical and literature results. 171

A Validation data for model prediction of *DR* in laminar flow

Table A.1: Drag reduction in channel flow due to wide surface structures: Comparison of the results presented in Figure 5.2 which are evaluated according to the model of Bahrami [9], numerical results and results from Pironneau and Arumugam [98].

$a/(a+b)$	l/L	**model result**	numerical result	result in literature [98]	l/b
1	9.4	**~50%**	–	50%	∞
1	4.2	**41%**	–	38%	∞
0.95	8.3	**~48%**	50%	–	100
0.91	4.2	**32%**	33%	–	25
0.74	8.3	**~28%**	30%	–	20
0.74	4.2	**18%**	8.5%	–	7.1
0.67	7.1	**22%**	23%	–	14
0.59	8.3	**~15%**	16%	–	14

Acknowledgements

First of all I would like to thank Prof. Dr.-Ing. Bettina Frohnapfel, who gave me the chance to work with her on the project "Innovative Oberflächenstrukturen zur Beeinflussung der Impulsübertragung zwischen Fluid und Feststoff". The financial support of the German Research Foundation (DFG) under grant number FR2823/2-1 is gratefully acknowledged.

I want to express my deep gratitude to Prof. Dr.-Ing. Bettina Frohnapfel and apl. Prof. Dr.-Ing. habil. Jovan Jovanović who were part of my education in fluid mechanics from the beginning and brought me into contact with the fascinating subject of turbulence. During my employment as a doctoral researcher they fostered the progress of my Ph.D. project by constant support, encouragement, inspiration and valuable advise.

Many thanks also to Dipl.-Ing. Andreas Güttler, Dipl.-Ing. Hermann Lienhart, M.Sc. Lukas Pflug and Dipl.-Ing. Alexander Stroh for the numerous fruitful discussions and exchange of perceptions in the context of transition, drag reduction and shape optimization research.

Moreover, I would like to thank all colleagues and students for the pleasant and productive time we spent together at the Institutes of Fluid Mechanics in Karlsruhe and Erlangen and at the Center of Smart Interfaces within the last few years. In particular, I want to thank Dipl.-Ing. Anna Slotosch for the enjoyable and motivating working atmosphere in our office and her collaboration and friendship. Further, I want to express my gratitude to M.Sc. Sebastian Türk, B.Sc. Veronika Krieger, Dipl.-Ing. Kolja Müller, Dipl.-Ing. Klaus Schmidt and M.Sc. Verena Fink for their huge commitment and effort in the preparation of their Bachelor, Master and diploma theses.

Finally, I would like to thank my family and friends for their understanding and motivating support. No words can express my gratefulness to Jonas for his constant and limitless confidence and encouragement. Thank you for your love and solidarity during the challenging years, which made a major contribution to this successful conclusion.